WITHDRAWN

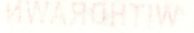

Computer Number Systems and Arithmetic

519.4
Sco85c

Computer Number Systems and Arithmetic

NORMAN R. SCOTT
The University of Michigan

Prentice-Hall, Inc.
Englewood Cliffs, New Jersey 07623

Library of Congress Cataloging in Publication Data

Scott, Norman R. (Norman Ross)   (date)
  Computer number systems and arithmetic.

  Includes bibliographical references and index.
  1. Computer arithmetic.   2. Numbers, Theory of.
I. Title.
QA76.9.C62S38   1984   519.4   84-9964
ISBN 0-13-164211-1

Editorial/production supervision and interior design: **Aliza Greenblatt**

Cover design: **Lundgren Graphics, Ltd.**

Manufacturing buyer: **Gordon Osbourne**

© **1985 by Prentice-Hall, Inc., Englewood Cliffs, New Jersey 07632**

*All rights reserved. No part of this book may be
reproduced, in any form or by any means,
without permission in writing from the publisher.*

Printed in the United States of America
10 9 8 7 6 5 4 3 2 1

ISBN 0-13-164211-1 01

PRENTICE-HALL INTERNATIONAL, INC., *London*
PRENTICE-HALL OF AUSTRALIA PTY. LIMITED, *Sydney*
EDITORA PRENTICE-HALL DO BRASIL, LTDA., *Rio de Janeiro*
PRENTICE-HALL CANADA INC., *Toronto*
PRENTICE-HALL OF INDIA PRIVATE LIMITED, *New Delhi*
PRENTICE-HALL OF JAPAN, INC., *Tokyo*
PRENTICE-HALL OF SOUTHEAST ASIA PTE. LTD., *Singapore*
WHITEHALL BOOKS LIMITED, *Wellington, New Zealand*

Contents

CATMar4'85

2-5-85 Pub. 31.96 Alden

84-4403

Preface

This book is an introductory treatment intended both for the newcomer to the subject and for the person experienced with present-day number systems and arithmetic who wishes a presentation of the new directions in which computer arithmetic is going. It may seem odd to publish a book on computer arithmetic at a time when computers (microcomputers especially) are being used increasingly for nonarithmetic functions. Yet it is also true that the area of arithmetic applications is growing rapidly, and it is important that the well-informed computer engineer should know a great deal about computer number systems and arithmetic. The remarkable progress being made in the large-scale integration of circuits on single semiconductor chips makes the unthinkably complex circuit of yesterday an easily realizable component tomorrow, and algorithms or implementations that may have seemed impossible to construct with discrete components suddenly become attractive possibilities with very large scale integration (VLSI) circuit technology. It is thus appropriate to examine both the conventional ways of doing computer arithmetic and the unconventional methods that may find application in new designs.

This book is not the definite treatise on computer arithmetic. Such a work has not yet been published, although there are several active research workers in this field who ought to write it. It is also not a compendium of clever logic circuits for implementing this or that algorithm in an ingenious fashion. It is rather an attempt to point out the essentials of number systems and arithmetic processes, with logic diagrams shown where they clarify the discussion. It is assumed that the reader has at least minimal familiarity with both computer hardware and computer software. Although logic circuits and their implementations are not treated, the notions of "gate," "flip-flop," and "register" are essential hardware concepts. Similarly, software ideas such as the stored program concept, loops, branches, and program sequences are also essential.

Norman R. Scott

NOTATIONAL CONVENTIONS

$\lfloor x \rfloor$ Smallest integer not less than x

$\lceil x \rceil$ Greatest integer not greater than x

$\{x, y\}$ Set inclusion (the set whose elements are x and y)

$x \in S$ "Element x is a member of the set S"

$O[f(n)]$ "of the order of $f(n)$," that is, "is less than some constant times $f(n)$, as $n \to \infty$"

ABBREVIATIONS FOR NAMES OF JOURNALS

1. *IEEE Trans. Comp*   *IEEE Transactions on Computers*
   IEEE Trans. El. Comp.   *IEEE Transactions on Electronic Computers*
   IRE Trans. El. Comp.   *IRE Transactions on Electronic Computers*

The above are all the same periodical, published initially by the IRE (Institute of Radio Engineers) and subsequently by the successor organization IEEE (Institute of Electrical and Electronics Engineers). The restrictive adjective "Electronic" was dropped from the title in 1968.

2. *Proc. IRE*   *Proceedings of the IRE*
   Proc. IEEE   *Proceedings of the IEEE*
3. *Proc. IEE*   *Proceedings of the Institution of Electrical Engineers* (a British professional society distinct from the IEEE)
4. *IBM Jl. R&D*   *IBM Journal of Research and Development* (published by International Business Machines Corp., Armonk, N.Y.

Computer Number Systems and Arithmetic

About Numbers

1.1 A BIT OF HISTORY

The origins of man's concept of number are hidden from us in the obscure prehistoric past and will never be fully known. Whatever speculations we may make about the way that early man may have gradually formulated the idea of number as a characteristic separate from particular numbers of particular objects, they remain speculations. Our guesses about this prehistory are made at least reasonable by the fact that there still exist certain isolated tribal groups (usually described in newspaper accounts as having a "stone age" culture) in which only a few, if any, number names exist and in which separate and distinct names are used for such different entities as "two fish," "two men," or "two spears." In these languages no word exists for the abstract concept "two," and we may well suspect that prehistoric man once struggled with similarly limited languages. If the language does not include a means for naming numbers, it seems self-evident that it cannot include descriptive terms for arithmetic processes, that is, for the processes of generating the numbers implied by operations on given numbers.*

The first records we have of the use of numbers date from the ancient Sumerian civilization of Mesopotamia. The Sumerians and their successors, the Babylonians, left tens of thousands of baked clay tablets inscribed with records of commercial transactions in the period 4000–1200 B.C. These tablets do not reveal anything about the de-

---

*The extent to which language influences thought processes has been widely examined by linguists and psychologists since the American linguist Benjamin Lee Whorf made his classic studies of the Hopi language in the 1920s and concluded that one's language is a principal factor in determining how one thinks. That is, how we think about things is determined by what our language enables us to say about them. Although the Whorfian thesis has not been universally accepted, it does appear to apply in the case of numerical reasoning.

velopment of the number concept, since they display a rich, well-developed, and widely used number system, one already in full flower rather than one struggling for definition. Although we might have expected this first recorded number system to be only a rudimentary one, it is surprisingly well developed and was the instrument of an advanced system of commerce. Like our present-day decimal system, the Sumerian system was a positional and based system; that is, digit position indicated what power of the base was to be multiplied by the digit value in evaluating the number. Unlike our system, the base was 60, and furthermore there was no symbol for zero. When no units in a particular position were to be specified, the ancient scribes and readers of records had to rely on contextual aids to indicate the positional significance of the other digits. It is now fairly well agreed that our symbol for zero came from India, reaching the Arabic world around the ninth or tenth century A.D., and finding its way thence to the Western world. (It is interesting to note the great debt owed by the Western world to the Moslem world, for at the very time that Europe was struggling through the centuries of the Dark Ages, Moslem culture was in full bloom, and the Moslem world was a repository of civilization.) Use of the symbol "zero" made possible much simpler representations of numbers than did other systems, such as Roman numerals.

Another feature of modern number systems that was not present in ancient times (or even in fairly recent times) was the concept of negative numbers. Although Bell, in his fascinating book on the history of mathematics [1], cites evidence that the Babylonians accepted negative numbers as well as positive numbers in solving pairs of simultaneous equations, the idea apparently did not take hold. Bell remarks: "The one glimmer of mathematical intelligence in the early history of negatives is the suggestion of Fibonacci that a negative sum of money may be interpreted as a loss." Since Fibonacci lived in the period around A.D. 1200, we can see how recently the concept of negative numbers has come into use.

1.2 NAMES FOR NUMBERS

We have already noted that primitive languages exist that have few words or symbols for representing numbers. On the other hand, the diversity of number names and symbols among various languages and cultures is very large. For example, the number that in English is called "nine" is recognized by other cultures and other ages by such names or symbols as

$$9 \quad\quad IX \quad\quad neun \quad\quad 九 \quad\quad \text{卅 IIII} \quad\quad ДеВЯТЬ$$

and many others. Yet each culture tends to limit its set of number names to those that are frequently needed in the daily affairs of its society. A society of sheepherders has no occasion to need the word "million," and "billion" and its successors have come into use in the English language only within the last century. "Trillion" finds use nowadays only in measuring that ill-defined quantity called the gross national product, although it seems likely that within a few more years Americans will learn to use it when speaking of the annual federal budget. "Quadrillion" and similarly formed names for bigger numbers begin to grow clumsy, and scientists and engineers prefer to use scientific notation, which uses exponents of 10 to indicate magnitudes.

We should note at the start that merely assigning names to numbers is not sufficient to define a number system. Since the number of numbers is infinite, we cannot arbitrarily assign names to all numbers but must use a recursive definition in which the names of most numbers will be expressed as combinations of names of other numbers. For example, the number name "six hundred" uses the names of two other numbers. Of course, the names themselves are usually used only in spoken communication, and for writing we use symbols of various kinds. In the computer we use yet other representations, such as voltage levels, states of magnetization, or combinations of holes or no holes.

It is interesting to observe that in English the unique number names grow sparser as the numbers grow bigger, and that to name larger numbers we are increasingly dependent on the recursive use of names of smaller numbers. Thus the first 21 numbers, 0 through 20, have their own individual names (although some of them originated as compound names). To name 21 through 29, we use combinations of other names. Then comes another unique name, "thirty" (again originally a compound word). Every tenth name beyond has a unique name up through "hundred." Then follow only compound names up to "thousand," again only compound names up to "million," and so on. As is so often the case in both natural and artificial languages, we find short symbols being used for frequently occurring "messages."

The distinguished Argentinian writer Jorge Borges, many of whose writings display his fascination with numbers, describes in the story "Funes the Memorious" [2] a strange "system" of number names invented by the mystic Funes:

> He told me that in 1886 he had invented an original system of numbering and that in a very few days he had gone beyond the twenty-four-thousand mark. He had not written it down, since anything he thought of once would never be lost to him. His first stimulus was, I think, his discomfort at the fact that the famous thirty-three gauchos of Uruguayan history should require two signs and two words, in place of a single word and a single sign. He then applied this absurd principle to the other numbers. In place of seven thousand thirteen, he would say (for example) *Maximo Perez*; in place of seven thousand fourteen, *The Railroad*; other names were *Luis Melian Lafinur, Olimar, sulphur, the reins, the whale, the gas, the caldron, Napoleon, Agustin de Vedia*. In place of five hundred, he would say *nine*. Each word had a particular sign, a kind of mark; the last in the series were very complicated . . . I tried to explain to him that this rhapsody of incoherent terms was precisely the opposite of a system of numbers. I told him that saying 365 meant saying three hundreds, six tens, five ones, an analysis which is not found in the "numbers" *The Negro Timoteo* or *meat blanket*. Funes did not understand me or refused to understand me.*

*Jorge Luis Borges, LABYRINTHS. Copyright © 1962 by New Directions Publishing Corporation, Reprinted by permission of New Directions.

We may note also that since earliest times numbers have possessed a mystical significance for some people. For example, the Roman writer Varro (116–27 B.C.) felt that "the virtues and powers of the number 7 are many and various," and therefore wrote 70 times 7 books and a set of 700 biographies [3]. Countless other examples of this nonsense exist, which is harmless in such innocent forms as omitting 13 when numbering the floors of hotels or office buildings but can become dangerous anti-intellectualism when it becomes popularized in such forms as the present-day fad of astrology.

1.3 HOW BIG ARE NUMBERS?

Let us examine for a moment the matter of how big numbers have to be to encompass the things we want to state numerically. Archimedes set the style for such investigations in his work "The Sand Reckoner" [4] by setting out to compute how many of the smallest elements known would be needed to fill up the biggest space known, choosing for the smallest element a grain of sand and for the largest space the universe as then known. The number involved was so far beyond the size of any needed in daily life of that time (the third century B.C.) that Archimedes was obliged to invent a new way of expressing very large numbers. He started with the Greek number "myriad" = 10,000, and noting that a myriad myriads can be expressed, he termed all lesser numbers "numbers of the first order." Then taking 10^8 as the unit size for the next range, numbers from that size to $(10^8)^2$ were termed "numbers of the second order." He similarly defined the third, fourth, and fifth orders, up through the 10^8 order, with the number $P = 10^{8 \times 10^8}$ being the largest number of this first period of numbers. The second period started from P, its first order extending to $10^8 P$, its second order to $(10^8)^2 P$, and its 10^8 order ending with $10^{8 \times 10^8} P$, that is, P^2. Continuing the process to the 10^8 order of the 10^8 period reaches at last the truly big number P^{10^8}, that is, $(10^{8 \times 10^8})^{10^8} = 10^{8 \times 10^{16}} =$ the digit 1 followed by 8×10^{16} zeros. Using this notation, he then went on to calculate the answer to his original problem, finding the number of grains of sand that could be encompassed by the universe to be less than 1000 units of the seventh order of numbers, that is, less than 10^{52}.

A more recent exploration of big numbers was described by the American mathematician Warren Weaver in an article in the *Atlantic Monthly* [5]. By this time, both the smallest particles and the universe were better defined. Weaver calculated the ratio of the diameter of the universe to the diameter of the proton to be about 10^{42}, a very big number by comparison with those that most of us meet in our daily affairs, but mathematically almost pitifully small. Weaver similarly found the age of the universe, when expressed in terms of the smallest meaningful time unit (which he took as 10^{-30} second, the period of the highest-energy cosmic rays known), to be 10^{47}, another big but not inconceivable number. His candidate for "biggest meaningful number" was one that he attributes to "an English mathematician named Skewes [who] in connection with a theorem about prime numbers found specific use for what we will call the Skewes number, namely:

$$10^{10^{10^{34}}}$$

This symbol stands for the utterly fantastic number which, if written out in full, would have

$$10^{10^{34}} = 10^{10,000,000,000,000,000,000,000,000,000,000,000}$$

zeros."

The latest such exploration was one conducted by D. E. Knuth, that Renaissance Man of computer mathematics, and reported by him in *Science* magazine for December 17, 1976 [6]. Knuth also briefly investigates some "astronomically" large numbers, but like Weaver, finds that the truly "big" numbers arise in problems in number theory and in particular those that involve combinatorial analysis. Following Archimedes' lead, Knuth also finds it expedient to introduce a notation for expressing large numbers. As are all good notations, his is both concise and simple:

1. We let $x \uparrow n$ denote n factors x multiplied together (which we ordinarily write as x^n).

2. By simple extension of this notation, $x \uparrow\uparrow n$ denotes $x \uparrow (x \uparrow (x \uparrow (\cdots \uparrow x) \cdots))$, where we take powers n times.

3. The general rule is

$$\underbrace{x \uparrow\uparrow \cdots \uparrow n}_{k \text{ arrows}} = \underbrace{\underbrace{x \uparrow\uparrow \cdots \uparrow}_{k-1} (x \underbrace{\uparrow\uparrow \cdots \uparrow}_{k-1} (x \underbrace{\uparrow\uparrow \cdots \uparrow}_{k-1} (\cdots \underbrace{\uparrow\uparrow \cdots \uparrow}_{k-1} x)) \cdots)}_{n \text{ times}}$$

Like most number system definitions, this one also is recursive, in that each group of arrows is defined in terms of groups having one fewer arrows, down to the terminating level $x \uparrow n = x^n$. It is also much more concise than the Archimedian notation, and it is interesting to note that Archimedes' biggest number is much smaller than the number $10 \uparrow\uparrow 2$. In a similar manner, even the "utterly fantastic" Skewes number is much smaller than $10 \uparrow\uparrow 4$, which is a very small number in Knuth's notation.*

Of course, it is not the purpose of the preceding discussion to demonstrate that computer hardware should be capable of representing enormously big numbers. Whatever the number range the designer elects to build into the hardware, the programmer has the liberty by suitable choice of scaling factors to use that number range for the representation of other number ranges. One part of our purpose is simply to demon-

---

*Indeed, a vastly bigger number than the Skewes number has been found by Ronald Graham of Bell Laboratories, as an upper bound on a certain combinatorial problem. It is described in *Science*, vol. 218, Nov. 19, 1982, p. 780.

strate that there are enormously many numbers "out there" to be manipulated by our computers, and the system designer is confronted by a very real challenge in trying to devise computer number representations and arithmetic systems that will be suitable for the ones most frequently encountered and at least usable on the occasional exceptional number.

1.4 THE USES AND MISUSES OF NUMBERS

Numbers play an enormously important role in today's advanced technological societies, which could not exist without them. We use numbers to express units of trade and commerce, monetary values, physical dimensions—indeed, to express the measure of any concept that can be quantified. One hundred years ago, the eminent scientist-engineer Lord Kelvin said: "In physical science a first essential step in the direction of learning any subject, is to find principles of numerical reckoning, and methods for practicably measuring, some quality connected with it. I often say that when you can measure what you are speaking about, and express it in numbers, you know something about it; but when you cannot measure it, when you cannot express it in numbers, your knowledge is of a meagre and unsatisfactory kind: it may be the beginning of knowledge, but you have scarcely in your thoughts, advanced to the stage of *science*, whatever the matter may be" [7].

The need to quantify and to assign numerical values has been carried so far that we are awash every day in a sea of numbers—weather reports, baseball statistics, election returns, stock market reports, grocery prices, calorie counts, census data, and a thousand other aspects of today's world. Indeed, many of the concepts to which we attach numbers are either intrinsically not quantifiable, or are precise concepts that are not widely understood, or are statistical concepts that are neither well defined nor widely understood. The reader has only to examine a daily newspaper to encounter references to batting averages, the consumer price index, the Dow Jones average, the gross national product, and public opinion polls predicting election outcomes, among other numerical measures, and it is an unusual reader who knows how all these numbers are calculated, what they signify, and how trustworthy they are. Lacking understanding of the source and meaning of these numerical quantities, most people are content to accept them reverently and to cite them glibly when the occasion arises. The urge to quantify and to number so dominates modern conversation and writing that no topic seems adequately treated without some numerical reference, and the most significant aspect of a great work of art may be the price it fetches at auction.

Although our entire culture blithely bandies numbers about, there is a widespread ignorance of numerical magnitudes and an equally broad incompetence at simple arithmetic. It has often been remarked that skill at arithmetic is unnecessary now that everyone has two or three pocket calculators, but the mere existence of a device to calculate does not imply that a user will know how to employ it, or what to do if the batteries wear out. The inability to understand letters, that is, to read, is known as "illiteracy," and we may similarly speak of "innumeracy"* to describe the inability to use numbers. This condition is far more general than might be suspected by computer people who talk only to computer people, and in today's technological world, there should be as much concern over "numeracy" as over literacy. The "two cultures" may prove to be those who can and those who cannot use numbers.

1.5 DEFINITION OF "NUMBER": PEANO'S AXIOMS

In the foregoing pages we have examined some of the historical and cultural aspects of numbers and the concept of "number" without attempting to define just what "number" might be. For most purposes of ordinary daily life, "number" is just another commonly accepted word of our customary language and is well understood by all of us without any more need for precise definition than any other word, such as "other" or "word." However, as we begin to investigate algorithms for manipulating numbers and hardware structures for implementing these algorithms, it is essential to have a clear understanding of what we mean by "number." This clear understanding is not at all facilitated by the fact that in the English language (at least), there is no clear distinction between "number" as an abstract concept and "number" as a set of symbols or a name for a particular instance of this concept. Thus "23" is said to be a *number*, and "twenty-three" is the name of that number. However, in other languages and other cultures, other symbols and other names are used. Behind the various symbols and names is the concept of "number," which is shared by all of them.

To define this abstract notion of "number," we make use of *Peano's axioms* defining the positive integers, that is, the set of "natural numbers":

The positive integers are the set of elements that satisfy the following postulates:
1. There is a positive integer 1 in the set.
2. Every positive integer a has a distinct successor positive integer a^+.
3. The element 1 is not the successor of any positive integer.
4. If two elements have the same successor, then they are the same element. That is, if $a^+ = b^+$, then $a = b$.
5. Every set of positive integers that contains 1 and the successor of every element of the set contains all the positive integers.

We may note several points about this set of axioms:

1. As do all axiom sets, it contains some undefined words and terms, for whose meaning we rely on the commonly accepted interpretations.

---

*I wish I had invented this felicitous term, but credit goes to an English writer (whose name I don't recall), writing in the periodical *New Society* sometime in the 1960s.

2. Although we have used the symbol "1," it has no significance except that given to it by the axioms. Any other mark would have served as well.

3. The important concept of "successor" corresponds closely to our intuitive understanding of that word (although we could have used another term, such as "correspondent"). It is this successor relationship that is the fundamental concept in our number representation systems, since we need rules for formulating the symbols that are successors to symbols.

4. The set of positive integers is an infinite set, since every element has a distinct successor.

5. The axioms do not address the question of structures of symbols to represent numbers, but rather the more fundamental idea of what "number" itself is.

The integers comprise a larger set than the positive integers defined by the Peano axioms. In order to solve such an equation as $a + x = b$ for any combination whatsoever of positive integers a and b, we define the set of integers to consist of

1. The set of positive integers previously defined, whose elements we denote by 1, 2, 3, . . . in the conventional manner.

2. The element zero

3. The negative integers, denoted by $-1, -2, -3, . . .$

That is, the set of integers consists of all those elements x that satisfy $a + x = b$ when a and b are any positive integers. Of course, this definition begs the question of the meaning of "+," that is, addition. For that, however, we turn to the concept of "successor," and we define:

$$x + 1 = x^+ \qquad \text{for all } x$$

and

$$x + y^+ = (x + y)^+ \qquad \text{for all } x \text{ and } y$$

where the superscript $+$ denotes the successor element, and the operator $+$ denotes addition. As was the case with the Peano axioms, this definition is independent of the particular representation we give to numbers either verbally, or with pencil and paper, or with aggregates of voltages and devices in a computer. The principal problem of computer arithmetic is to find representations for numbers that lend themselves readily to simple devices and procedures for ascertaining this successor relationship.

REFERENCES

1. Bell, Eric Temple, *The Development of Mathematics*, McGraw-Hill, New York, 1945, p. 34.
2. Borges, Jorge, *Labyrinths*, New Directions, New York, 1964, pp. 59–66.

3. *Encyclopaedia Britannica*, 1960 Edition, vol. 22, Encyclopaedia Britannica, Chicago, p. 945.

4. Archimedes, The sand reckoner, in *The World of Mathematics*, vol. 1, ed. J. R. Newman, Simon and Schuster, New York, 1956, pp. 420–429.

5. Weaver, Warren, Size, *Atlantic Monthly*, Sept. 1948, pp. 88–90.

6. Knuth, Donald E., Mathematics and computer science: coping with finiteness, *Science*, vol. 294, no. 4271, Dec. 17, 1976, pp. 1235–1242.

7. Lord Kelvin (Sir William Thomson), *Popular Lectures and Addresses*, vol. 1, Macmillan, London, 1889, pp. 72–73.

Decimal Numbers

2.1 DEFINITIONS

As we begin to consider in careful detail the various possible ways of using numerical symbols to represent numbers, it is reasonable to start by examining the familiar system of numeration used in everyday life. In this system, any number can be written as a string of digit symbols chosen from the set

$$S = \{0, 1, 2, 3, 4, 5, 6, 7, 8, 9\}$$

(We do not attempt to define the meanings of these various symbols and regard them as having the commonly accepted interpretation, that is, "zero," "one," etc.) We use the symbol d, with appropriate subscript, to denote each of the digits in such a string. For example, we might designate the digits of an n-digit number as

$$d_{n-1}d_{n-2}d_{n-3} \cdots d_1 d_0$$

We commonly write

$$N = d_{n-1}d_{n-2}d_{n-3} \cdots d_1 d_0$$

(e.g., $N = 748$). Observe that the string of digits has an interpretation as a number only by virtue of commonly accepted understanding as to what the positions of the string elements imply. Thus "748" implies a number whose value is found by adding seven times one hundred, four times ten, and eight (times one). Changing the order of the string (e.g., to 874) gives a representation for a very different number, even though the same digits are used.

Our familiar decimal number system is a *positional weighted* system, in which a weighting factor is associated with the digit in each position. Further, it is a *based* number system, in that each weighting factor is a power of a base value (in this case

10), and it uses a *fixed* base, that is, the same base for all digit positions. Thus an integer $N = d_{n-1} d_{n-2} \cdots d_1 d_0$ is interpreted as meaning

$$N = d_{n-1} \times 10^{n-1} + d_{n-2} \times 10^{n-2} + \cdots d_1 \times 10^1 + d_0 \times 10^0$$

Note that if the weighting factors are written out explicitly, as in this example, we never need to use the digit symbol "0," since zero multiples of any weighting factor can be indicated by merely omitting mention of that weighting factor in the explicit sum form of N. However, if N is indicated in the more usual form of a string of digits, without explicit inclusion of the weighting factors, then the implicit weighting factors are indicated merely by the positions of the digits, and we must explicitly write a "0" in a position where no units of that weight are involved. Thus

$$7 \times 10^2 + 0 \times 10^1 + 3 \times 10^0 = 7 \times 10^2 + 3 \times 10^0$$

but $7 \ 0 \ 3 \neq 73$.

By way of contrast with our decimal system, consider the Roman numeral system. Here the digit symbols are chosen from the set

$$R = \{I, V, X, L, C, D, M\}$$

the symbols representing respectively 1, 5, 10, 50, 100, 500, and 1000. Numerals are read from left to right, the weights of the various units being added, except that any units that precede larger units are subtracted from the total. Thus

$$IV \longleftrightarrow 4$$
$$VI \longleftrightarrow 6$$
$$IX \longleftrightarrow 9$$
$$MC \longleftrightarrow 1100$$
$$CM \longleftrightarrow 900$$
$$MCMLXXVIII \longleftrightarrow 1978$$

Observe both the absence of zero and the unusual positional relation. Not only is this system inconvenient for the writing of numbers, but it is extremely awkward for such operations as multiplication and division. The only modern uses of the Roman numeral system are for inscriptions on cornerstones (and this is a dying art form) or as counterexamples for budding computer engineers.

Although our first example of a fixed-base number system used 10 as the base, any other base (except 1 or 0) can also be used. As a practical matter, we are interested only in bases that are whole numbers, and for the present we will discuss only bases that are positive whole numbers. (Negative bases are also possible, as we shall see shortly.) A base r number system requires r digit symbols, and numbers are written as strings of digits chosen from this symbol set. Each digit states the number of units of its implicitly associated weighting factor that are to be included in the number, and as in the decimal case, the weighting factors are integer powers of the base. The powers, of course, can be negative as well as positive, and thus our number system can represent fractions as well as integers. It is our modern custom to write mixed numbers as a string of digits with a period intervening at the boundary between that part of the string representing the integer portion and the part representing the fraction. In a base 10 system, this period is called the *decimal point*, although the more general name is *radix point*, when the base (or *radix*, as it is also called) is other than 10.

What symbols should we use for the digits in the general case where the radix $r \neq 10$? Although it is certainly possible to invent new symbols, the usual practice is to use our familiar set

$$S = \{0, 1, 2, \ldots, 9\}$$

truncating it if $r < 10$ or augmenting it with additional symbols if $r > 10$. For example, if we choose $r = 8$, to define the familiar octal number system, then

$$S_8 = \{0, 1, 2, \ldots, 7\}$$

Since these symbols are also used to represent decimal numbers, a digit string chosen from this set cannot be distinguished from the same string chosen from the set of decimal symbols. For example, 347 could be interpreted either as

$$3 \times 10^2 + 4 \times 10^1 + 7 \times 10^0$$

or as

$$3 \times 8^2 + 4 \times 8^1 + 7 \times 8^0$$

This possibility of confusion arises from our decision to use the same symbol set, and we must therefore include an explicit indication of the radix, in situations where confusion might occur. We would therefore write 347_{10} or 347_8 to distinguish the two possible interpretations.

A commonly used radix larger than 10 is the radix 16, and to use it we must augment our 10 digit symbols by six more. Some years ago, early in the electronic computer era, it was proposed that a wholly new set of symbols should be devised, having appearances suggestive of the numbers 10, 11, and so on, which they were to represent. This proposal received little support, and computer people very early settled on the practice of using the capital letters A through F to represent the numbers 10 through 15 in base 16 representations. For example,

$$2 A E 3_{16} = 2 \times 16^3 + 10 \times 16^2 + 14 \times 16^1 + 3 \times 16^0$$

and the reader may easily determine the decimal number that is equivalent.

2.2 RADIX CONVERSION

A common problem in computer work is the conversion of a number from one radix to another; that is, given a number N expressed as a digit string based on radix r_S (the radix of the source), we seek the digit string representing the same number expressed in terms of radix r_D (the radix of the destination). As we have already seen, N may be regarded as

$$N = d_{n-1} r_D^{n-1} + d_{n-2} r_D^{n-2} + \cdots + d_1 r_D^1 + d_0 r_D^0$$

where the digits d_j are the as yet unknown digits of N when r_D is used as the radix. To determine these digits we note that N may be written in nested polynomial form as

$$N = \{[(d_{n-1} \times r_D + d_{n-2}) \times r_D + d_{n-3}] \times r_D + \cdots + d_1\} \times r_D + d_0$$

or, equivalently,

$$
\left.
\begin{aligned}
N = N_0 &= N_1 r_D + d_0 \\
N_1 &= N_2 r_D + d_1 \\
N_2 &= N_3 r_D + d_2 \\
&\;\;\vdots \\
N_{n-1} &= d_{n-1}
\end{aligned}
\right\}
\Longrightarrow
\left\{
\begin{aligned}
N &= N_1 r_D + d_0 \\
N_j &= N_{j+1} r_D + d_j \\
N_{n-1} &= d_{n-1}
\end{aligned}
\right.
$$

Since (by definition), $0 \le d_j < r_D$ for all j, we may find each d_j in turn as the remainder upon dividing N_j by r_D, where N_0 is the integer value to be converted and $N_{j+1} = \lfloor N_j/r_D \rfloor$. The arithmetic operations are carried out in the arithmetic system of radix r_S. The algorithm may be written as

Algorithm 2.1
1. Initialize.
   (a) Set $j = 0$.
   (b) Set $N_j = N$.
2. While $N_j \ne 0$ do
   (a) d_j = remainder of N_j/r_D.
   (b) $N_{j+1} = \lfloor N_j/r_D \rfloor$.
   (c) $j = j + 1$.

For example, to convert the number 109_{10} to binary form (i.e., base 2 form), we proceed as follows:

Initial $N = 109_{10}$	
Quotients after division by 2	Remainders after division by 2
54	$1 = d_0$
27	$0 = d_1$
13	$1 = d_2$
6	$1 = d_3$
3	$0 = d_4$
1	$1 = d_5$
0	$1 = d_6$

Thus we have $109_{10} = 1101101_2$.

A closely related procedure may be developed for conversion of fractions, for if N is a fraction, we can write it in terms of the new radix r_D as

$$N = d_{-1}r_D^{-1} + d_{-2}r_D^{-2} + \cdots + d_{-n}r_D^{-n}$$

We may find each d_{-j} in turn as the integer part of the product of r_D times N_{-j}, where N_{-1} is the fraction to be converted and $N_{-(j+1)}$ is the fraction part of $r_D \times N_{-j}$, again using the arithmetic of the source radix r_S. We may write the algorithm as follows, letting n be the number of digits desired in the result.

Algorithm 2.2
1. Initialize.
   (a) Set $j = 1$.
   (b) Set $N_{-j} = N$.
2. While $j \neq n + 1$ do
   (a) $d_{-j} = \lfloor N_{-j} \times r_D \rfloor$.
   (b) $N_{-(j+1)} =$ fraction part of $N_{-j} \times r_D$.
   (c) $j = j + 1$.

For example, to convert the fraction $N = 0.703125_{10}$ to binary form, we proceed as follows:

| | Initial $N = 0.703125$ |
Integer part after multiplying fraction by 2	fraction part after multiplying fraction by 2
1	406250
0	81250
1	6250
1	250
0	50
1	0

Thus we have $0.703125_{10} = 0.101101_2$. Clearly, mixed numbers can be treated by executing these two algorithms in turn on their integer and fraction parts.

Several points about these procedures deserve special comment. It is clear that fractions that terminate (i.e., end with an infinite string of zeros) may not necessarily terminate when converted to another radix, simply because the bigger radix is not necessarily an integer power of the smaller radix. Our example above was deliberately selected to terminate when converted to base 2, but it is easy to choose examples that do not. Consider, for instance, $N = 0.1_{10}$, which, after execution of several iterations of our algorithm, becomes

$$N = (0.000110011001100\ldots)_2$$

Thus the conversion of fractions is executed only as many times as are needed to generate the desired number of digits in the new radix.

A second and somewhat more subtle point is that these two conversion processes must be executed using the arithmetic of the source radix. In the preceding examples, we started with numbers expressed in base 10 and used the familiar decimal arithmetic of everyday life. Suppose, however, that we wish to convert the octal (base 8) number $N = 735_8$ into decimal form. Our procedure requires us to divide repeatedly by ten, that is, by 10_{10}, but before we can generate the desired quotient and remainder digits, we should define an addition table and a table of multiples of the destination radix 10_{10}, all expressed in terms of radix 8 (the source radix). These appear below.

+	0	1	2	3	4	5	6	7
0	0	1	2	3	4	5	6	7
1	1	2	3	4	5	6	7	10
2	2	3	4	5	6	7	10	11
3	3	4	5	6	7	10	11	12
4	4	5	6	7	10	11	12	13
5	5	6	7	10	11	12	13	14
6	6	7	10	11	12	13	14	15
7	7	10	11	12	13	14	15	16

In base 10	In base 8
10	12
20	24
30	36
40	50
50	62
60	74
70	106
80	120
90	132

Now we may perform the division:

$$\text{(a)} \quad 12_8 \overline{)735_8} \quad \begin{array}{r} 57 \\ \underline{62} \\ 115 \\ \underline{106} \\ 7 \end{array}$$

$$\text{(b)} \quad 12_8 \overline{)57_8} \quad \begin{array}{r} 4 \\ \underline{50} \\ 7 \end{array}$$

$$\text{(c)} \quad 12_8 \overline{)4_8} \quad \begin{array}{r} 0 \end{array}$$

So $735_8 = 477_{10}$.

The reader has probably already observed that it is not necessary to use our more general algorithm to convert numbers from an arbitrary radix to base 10 form, since merely carrying out in decimal arithmetic the valuation of the polynomial implied by the source number will give its value in base 10. Thus

$$735_8 = 7 \times 8^2 + 3 \times 8^1 + 5 \times 8^0$$

$$= 7 \times 64 + 3 \times 8 + 5$$

$$= 477$$

However, using the arithmetic of the source radix r_S in converting to radix r_D is a direct method when neither r_S nor r_D is equal to 10, since it avoids the cumbersome conversion from r_S to base 10 and thence to r_D. As an exercise, it is instructive to set up whatever tables are needed to assist in converting from base 8 to base 7, and from base 7 to base 8, without an intervening conversion to base 10, and then to verify that the following equivalences are true:

$$735_8 = 1251_7$$
$$2303_7 = 1504_8$$

An easy-to-remember procedure for converting binary integers to decimal form, based on direct evaluation of the polynomial, is the following:

> Scan the binary number from the left, keeping a running sum S, which is zero until the leftmost 1 digit is encountered. Set $S = 1$, and thereafter, as each digit d_j is scanned, set $S = 2S + d_j$, until d_0 has been included.

It is easy to carry out this procedure mentally, for small strings of binary digits, and with obvious modifications it can be used also for converting binary fractions.

The two procedures that we have so far defined both use the arithmetic of the source radix; that is, in converting from r_S to r_D we use the arithmetic rules of r_S. We can also define procedures that use the arithmetic of r_D, the destination radix. All that is required is that the source digits and the source radix be expressed in terms of the destination radix. The corresponding procedures are then:

Algorithm 2.3   For conversion of integers, evaluate

$$N = [(d_{n-1} \times r_S + d_{n-2}) \times r_S + \cdots + d_1] \times r_S + d_0$$

Algorithm 2.4.   For conversion of fractions, evaluate

$$N = r_S^{-1} \times \{d_{-1} + r_S^{-1} \times [d_{-2} + \cdots + r_S^{-1} \times (d_{-(n-1)}) + r_S^{-1} \times d_{-n}] \cdots \}$$

2.2.1 Conversion Using Binary-Coded Decimal Digits

In practice, radix conversion within the computer usually involves only base 10 and base 2, with the decimal digits being represented individually in binary form. In the next section we examine in detail some of the many possible ways in which individual decimal digits can be represented by binary encodings, but for the remainder of this section the particular binary-coded-decimal-digit (BCDD) pattern that we will use is shown in Table 2.1.

TABLE 2.1

Decimal digit	Weight			
	2^3	2^2	2^1	2^0
0	0	0	0	0
1	0	0	0	1
2	0	0	1	0
3	0	0	1	1
4	0	1	0	0
5	0	1	0	1
6	0	1	1	0
7	0	1	1	1
8	1	0	0	0
9	1	0	0	1

Binary encodings

The need for radix conversion typically arises only in the following forms:

1. Conversion of integers from binary form to BCDD form
2. Conversion of fractions from binary form to BCDD form
3. Conversion of integers from BCDD form to binary form
4. Conversion of fractions from BCDD form to binary form

Conversions 1 and 2 are output problems, and conversions 3 and 4 occur at input. The four algorithms that we have defined can be applied directly to these four situations. (Of course, if numbers are manipulated within the computer in BCDD form, there is no need to perform radix conversion. Many computers allow the user the option of carrying out decimal arithmetic on the BCDD numbers or doing binary arithmetic on the binary numbers.)

One way of doing the radix conversion is to program the central processing unit (CPU) to execute the steps of the conversion algorithm, and another is to devise special-purpose hardware to do the job. If we program the conversion, we would use the following algorithms:

1. To convert integers from binary form to BCDD form, use Algorithm 2.1, with the destination radix $r_D = 10_{10}$ expressed in binary form as 1010. The remainder after each division step is a 4-bit group, and the successive remainders are the successively more significant BCD digits of the desired result. Observe that although multiplication by $\frac{1}{10}$ is equivalent to division by 10, it is not a satisfactory equivalent for two reasons:
   (a) The binary equivalent of $\frac{1}{10}$ cannot be expressed exactly.
   (b) The result of multiplying by $\frac{1}{10}$ forms an integer part (which is the same as the quotient in division by 10) and a fraction part that represents the desired

digit, but which must be converted to BCDD form by being multiplied by 10_{10}. Thus the process requires an extra multiplication, and one of the factors is inaccurate.

2. To convert fractions from binary form to BCDD form, use Algorithm 2.2, with the destination radix $r_D = 10_{10}$ expressed in binary form as 1010.

3. To convert integers from BCDD form to binary, use Algorithm 2.3, with the source radix $r_S = 10_{10}$ expressed in binary form as 1010.

4. To convert fractions from BCDD form to binary, use Algorithm 2.4, with the source radix $r_S = 10_{10}$ expressed in binary form as 1010.

2.2.2 Hardware Structures for Radix Conversion

In some situations, however, we may prefer to relieve the CPU of the burden of radix conversion computations by providing hardware to do the job. Circuit arrangements can be either serial (one binary digit is converted at each clock pulse), or parallel (all the binary digits are accepted or formed simultaneously) or some intermediate form between these two. The serial circuits are simpler but slower.

The serial techniques are based upon a procedure first described by J. F. Couleur [1]. We construct two sets of registers, as shown in Fig. 2.1. Each of the 4-bit stages can be shifted to the right, with its rightmost bit entering at the left end of the stage to the right. The rightmost stage shifts into the simple binary register. We assume also logic circuitry for modifying the contents of the BCDD stages. The algorithm for BCDD integer to binary form is as follows:

Algorithm 2.5.   BCDD integer to binary, with the BCDD number initially stored in the BCDD stages (least significant digit at the right)

1. Shift the entire bit pattern (BCDD stages and binary register) one position to the right.

2. Subtract 3 from any BCDD position that now holds a value ≥ 8.

3. Repeat steps 1 and 2 n times (where n is the length of the binary register).

This procedure divides the decimal number by 2 at each step, putting the remainder into the binary register. Since a digit at the low end of a BCDD stage represents a unit of weight 10 times as great as the weight of the stage to its right, division by 2 should put 5 units into the stage to the right. Since the right shift puts a unit into the 8's column of the next lower stage, we must subtract 3 to get the desired increment

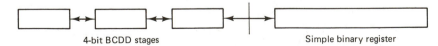

4-bit BCDD stages   Simple binary register

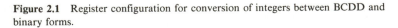

Figure 2.1   Register configuration for conversion of integers between BCDD and binary forms.

TABLE 2.2

BCDD form of 55

5 ↖   ↘ 5

Initially:	0 1 0 1	0 1 0 1	0 0 0 0 0 0
1. Shift	0 0 1 0	1 0 1 0	1 0 0 0 0 0
Subtract 3	0 0 1 0	0 1 1 1	1 0 0 0 0 0
2. Shift	0 0 0 1	0 0 1 1	1 1 0 0 0 0
No subtraction needed			
3. Shift	0 0 0 0	1 0 0 1	1 1 1 0 0 0
Subtract 3	0 0 0 0	0 1 1 0	1 1 1 0 0 0
4. Shift	0 0 0 0	0 0 1 1	0 1 1 1 0 0
No subtraction needed			
5. Shift	0 0 0 0	0 0 0 1	1 0 1 1 1 0
No subtraction needed			
6. Shift	0 0 0 0	0 0 0 0	1 1 0 1 1 1

Binary equivalent
of 55_{10}

of 5. For example, consider the number 55_{10} to be initially stored in two BCDD stages. The steps that convert this number are shown in Table 2.2

The process can be reversed to allow us to convert binary integers to BCDD form, if we allow the registers to be shifted to the left and provide slightly different adjustment of the BCD digits. The algorithm is:

Algorithm 2.6.   Binary to BCDD integer form
1. Add 3 to any BCDD group that holds a value of 5 or more.
2. Shift the entire bit pattern one position to the left.
3. Repeat steps 1 and 2 n times, to empty the n-bit binary register.

Quite similar procedures can be used for fractions, but now we put the binary register on the left and the BCDD stages on the right as in Fig. 2.2. The algorithms are:

Algorithm 2.7.   BCDD fractions to binary fractions
1. Add 3 to each BCDD group ≥ 5.
2. Shift the entire pattern one position left.
3. Repeat steps 1 and 2 n times to fill the n-stage binary register.

Algorithm 2.8.   Binary fractions to BCDD fractions
1. Shift the entire pattern one position right.
2. Subtract 3 from any BCDD group ≥ 8.
3. Repeat steps 1 and 2 n times to empty the n-stage binary register.

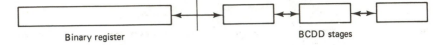

Figure 2.2 Register configuration for conversion of fractions between BCDD and binary forms.

In constructing circuit arrangements to do these tasks, we would not perform an explicit addition or subtraction of 3 but would instead arrange for each BCDD stage to go immediately to its correct next state, based on its present state and the digit being shifted into that stage. For example, Algorithm 2.6 is equivalent to the state table shown in Table 2.3.

From this table, it is a straightforward (if tedious) matter to derive the excitation conditions for the J and K inputs of each of the flip-flops and thus to arrive at the circuit of Fig. 2.3 for one of the BCDD stages.

We can derive a similar circuit for converting from BCDD integer to binary, at a rate that forms a new binary digit at every clock pulse until all the bits have been generated. However, a better circuit is possible, which accepts a BCD digit at each clock pulse and forms in the binary register the binary equivalent of all the decimal digits examined up to that time. We now associate the logic with the stages of the binary register and arrange for each stage to accept four inputs and generate four outputs. The next state of each stage and its four outputs will be determined by its present state and its four inputs. After the last decimal digit has entered the binary stages and all signals have propagated along the chain, the states of the stages $b_n b_{n-1} \cdots b_1 b_0$ represent the binary form of the original decimal number.

To carry out the process, we shift the BCD number to the left (i.e., most signifi-

TABLE 2.3

	If $I_{in} = 0$:		If $I_{in} = 1$:	
Present state	Next state	Stage output	Next state	Stage output
0 0 0 0	0 0 0 0	0	0 0 0 1	0
0 0 0 1	0 0 1 0	0	0 0 1 1	0
0 0 1 0	0 1 0 0	0	0 1 0 1	0
0 0 1 1	0 1 1 0	0	0 1 1 1	0
0 1 0 0	1 0 0 0	0	1 0 0 1	0
0 1 0 1	0 0 0 0	1	0 0 0 1	1
0 1 1 0	0 0 1 0	1	0 0 1 1	1
0 1 1 1	0 1 0 0	1	0 1 0 1	1
1 0 0 0	0 1 1 0	1	0 1 1 1	1
1 0 0 1	1 0 0 0	1	1 0 0 1	1

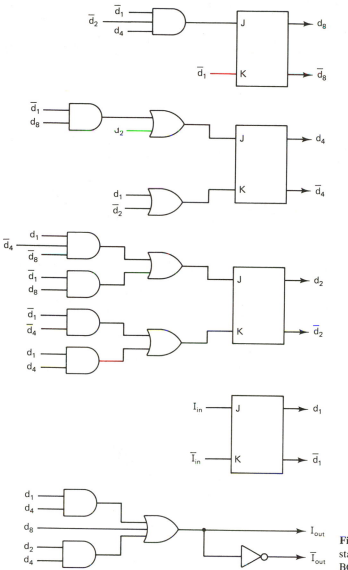

Figure 2.3 Logic structure of a BCDD stage for conversion of binary integers to BCDD form.

cant digit first) into the binary stages. Each binary stage assumes the state (1 or 0) of the low-order bit of the four that it receives and transmits to the left a new 4-bit group formed as follows:

1. If the state of the binary stage was 0 after the previous step, the 4-bit group received is shifted right by one position and is then transmitted to the left, or

Decimal input

	b6	b5	b4	b3	b2	b1	b0	1	0	9
Initially	0	0	0	0	0	0	0	0001	0000	1001
T₁ Transmitted values:	0000	0000	0000	0000	0000	0000	0001	0000	1001	----
States:	0	0	0	0	0	0	1	1001	----	----
T₂ Transmitted values:	0000	0000	0000	0001	0010	0101	0000	1001	----	----
States:	0	0	0	1	0	1	0	----	----	----
T₃ Transmitted values:	0001	0011	0110	0011	0111	0100	1001	----	----	----
States:	1	1	0	1	1	0	1	----	----	----

Binary result

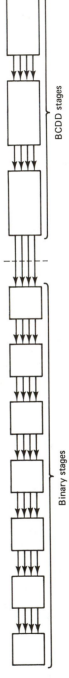

Binary stages   BCDD stages

Figure 2.4 Successive register contents in conversion of 109_{10} from BCDD to binary.

22

2. If the state of the binary stage was 1 after the previous step, the 4-bit group received is shifted right by one position and is then augmented by 5 (binary 0101). The resulting group is then transmitted to the left.

In each case, the right shifts are of the conventional type, in which 0's fill in vacated left positions.

Figure 2.4 shows a block diagram of the structure and a set of values occurring in the conversion of 109_{10} to its binary form 1101101_2. As before, we can exhibit this behavior in a state table (Table 2.4), where now (to simplify our notation) we show the inputs and outputs as decimal numbers rather than as 4-bit groups. Once again, routine "crank grinding" yields a circuit, shown in Fig. 2.5. Similar analyses permit design of circuits for converting the fraction forms, but we leave these to the reader to develop or to look up in Lanning's original description of these ideas [2].

The next step in circuitry for radix conversion is a completely parallel circuit in which all the binary digits of the input form are processed simultaneously, producing all the output digits in parallel (after time delays for signal propagation along the various paths through the structure). The fundamental ideas involved in using a rectangular array of identical elements for this purpose were presented by J.–D. Nicoud [3] and subsequently Bredeson [4] showed a very effective application of these ideas with the structure of Fig. 2.6. The iterated array consists of identical *doubler-halver* circuits, all of which are simultaneously placed in the doubling mode or in the halving mode by a

TABLE 2.4

Present state of binary stage	Input group	Output group	Next state
0	0	0	0
0	1	0	1
0	2	1	0
0	3	1	1
0	4	2	0
0	5	2	1
0	6	3	0
0	7	3	1
0	8	4	0
0	9	4	1
1	0	5	0
1	1	5	1
1	2	6	0
1	3	6	1
1	4	7	0
1	5	7	1
1	6	8	0
1	7	8	1
1	8	9	0
1	9	9	1

Input group: Outputs to next stage:

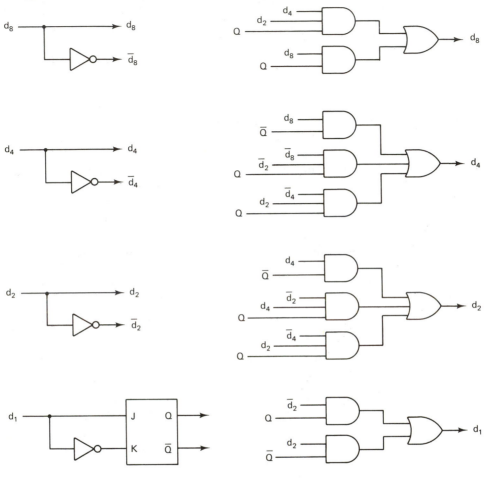

Figure 2.5   Logic structure of a BCDD stage for conversion of BCDD integers to binary form.

common external control signal (which is not shown, to simplify the diagram). The double arrows represent parallel 4-bit groups. When the conversion is from BCDD form to binary, the BCD digits are applied at the top and the binary equivalent digits emerge at the left side or the right side, depending on whether the input was a fraction or an integer. For binary to BCDD form, the inputs are applied at left or right, and the BCDD outputs are formed at the bottom.

The doubler-halver cell, which is the heart of the system, is a logic array that performs the conditional addition or subtraction of 3 that we noted in Couleur's algorithm. It has four inputs at the top and four outputs at the bottom, corresponding to the 4 bits of a BCD digit. As a doubler, it has a C_i bit in at the right and a C_o bit out at

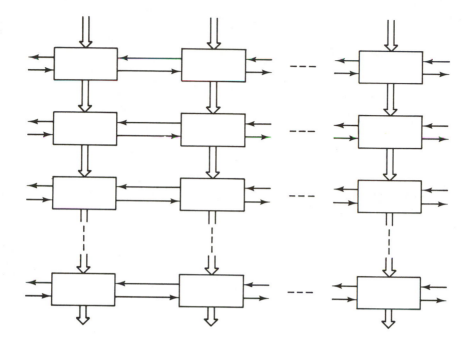

Figure 2.6 Array of 4-bit doubler-halver circuits for binary-to-decimal conversion.

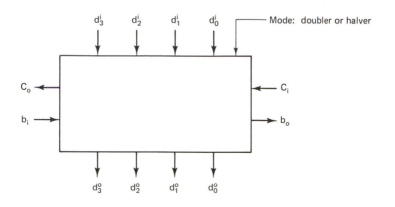

Figure 2.7 Input and output connections for the doubler-halver cell.

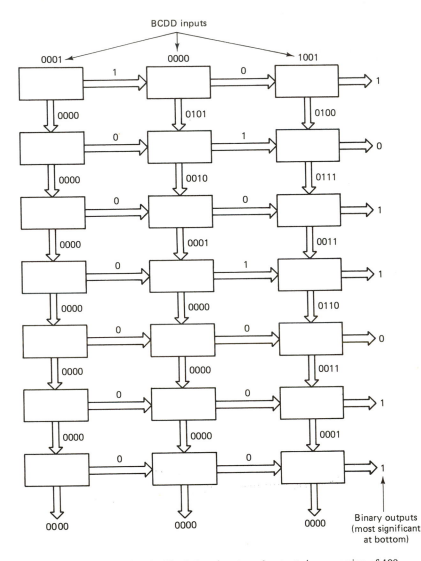

Figure 2.8 Successive doubler-halver inputs and outputs in conversion of 109_{10} from BCDD to binary.

the left, and as a halver it has a b_i bit in at the left and a b_o bit out at the right, as in Fig. 2.7. When the cell acts as a doubler, it performs an action equivalent to:

1. Receives a 4-bit input number and a C_i from the right.
2. If the sum of these $\geq$ 5, adds 3.
3. Shifts the result one bit position left, producing a one-bit C_o at the left and a 4-bit group $d_3^o d_2^o d_1^o d_0^o$.

When it is in the halving mode, its action is equivalent to:

1. Receives a 4-bit input number at the top and a b_i from the left.
2. Forms $b_o = d_0^i$.
3. If the digit group $b_i d_3^i d_2^i d_1^i$ represents a BCD digit ≥ 8, it subtracts 3 and sets the output equal to the result. Otherwise, no subtraction of 3 occurs.

The overall array can be operated in the following ways:

Conversion of:	Inputs	Outputs	Mode
1. BCDD fractions to binary fractions	Applied at top, most significant BCD digit at left	Produced at left, most significant bit at the top	Doubler
2. BCDD integers to binary integers	Applied at top, most significant BCD digit at left	Produced at right, most significant bit at the bottom	Halver
3. Binary integers to BCDD integers	Applied at right, most significant bit at the top	Produced at the bottom, most significant BCD digit at left	Doubler
4. Binary fractions to BCDD fractions	Applied at left, most significant bit at the bottom	Produced at the bottom, most significant BCD digit at left	Halver

In each case, the two unused sets of inputs are held at 0.

Figure 2.8 shows the digit patterns in a small array converting 109_{10} from its BCDD form to binary. For large numbers, this array can become so large as to be of doubtful practicality. For instance, conversion of 34-bit numbers requires an array which is 10×34, or 340 doubler-halver elements, and 68-bit numbers require four times this many. Benedek [5] has shown ways in which a portion of such an array can be used iteratively. Such a serial–parallel procedure can save equipment at the cost of time, without incurring the penalty of wholly serial operation.

2.3 BINARY ENCODINGS OF DECIMAL NUMBERS

All the internal operations of any digital system are carried on by means of *binary* signals, that is, signals that are restricted to being one or the other of only two allowed values. Some of these "signals" are intrinsically binary: for example, the presence or absence of a hole at a specified location on a punched card or a paper tape, and others, such as voltage levels at certain points, are deliberately designed to be binary. Electrically, it is entirely possible to let a given wire carry any one of 10 distinct voltage levels that can correspond to the 10 decimal digits, but it would be very difficult to ensure that each of the tens of thousands of elements in the computer could be designed to

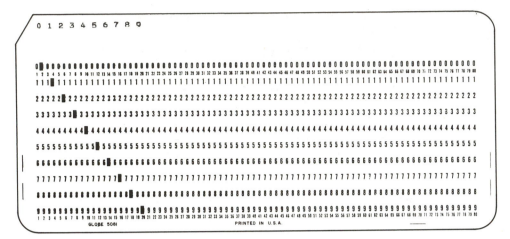

Figure 2.9 Punched card with hole patterns for 0, 1, . . . , 9.

have these same levels and that these levels would not variously drift up or down with time or temperature or other conditions. It has been found far simpler and more reliable to design devices that have only two allowed voltage levels. Thus from the moment a keypunch operator strikes an input key until the moment a print hammer strikes the output paper, all numbers (and indeed all characters) are represented in the machine by patterns of binary digits. Figure 2.9, for example, shows the various patterns of holes punched on a card for the various decimal digits 0 through 9. These configurations of holes or no holes represent one particular manner of "encoding" the decimal digits into binary configurations. Infinitely many ways can be devised to represent decimal digits by binary patterns, and we will examine here a few of these that are of interest to computer engineers.

What is the minimum number n of binary digits that we can use to represent a decimal number of m digits? Since the biggest number we can write with m decimal digits is a string of m 9s, which is equal to $10^m - 1$, we must choose n so that $2^n - 1 \geq 10^m - 1$; that is, we should take the smallest n such that $2^n \geq 10^m$. Thus

$$n = \left\lceil \frac{m}{\log_{10} 2} \right\rceil$$

or n is approximately $3\frac{1}{3} \times m$ rounded up to the next integer. For instance, 10-decimal-digit numbers require 34 digits when expressed in base 2 form.

By the same argument we see that a single decimal digit requires [3.32] binary digits—three are not enough, since they have only eight possible combinations, and four, while permitting the necessary 10 combinations, have six superfluous combinations. Four digits, of course, are what we must use at the minimum, since we cannot handle part of a digit (e.g., part of a hole or part of a no-hole).

Note that four binary digits can be arranged in 16 different combinations. If we make completely arbitrary assignments of these combinations to represent the decimal digits, we may select any one of the 16 to represent 0, any one of the remaining 15 to represent 1, and so on, defining in this way $16!/6! = 29,059,430,400$ different 4-bit encodings. There is a sense in which these encodings are not really all "different" since many can be derived from others by complementing digits (i.e., exchanging 1's and 0's) or by rearranging them. If the four binary digits are designated x_1, x_2, x_3, and x_4, we see that there are $4! = 24$ sequences in which they may be arranged. (For example, in Table 2.5, any one of the four columns of binary digits may be put on the left, any one of the remaining three next, and so on.) Moreover, any column or all of them may be complemented, giving (for each of the 24 sequences) 16 possible ways of distributing complementation symbols. Thus each encoding is one of a family of $16 \times 24 = 384$ codes that can be derived from one another, and we may consider the number of fundamentally different codes to be $(16!/6!) \times 1/_{384} \simeq 7.6 \times 10^7$. Out of this enormous number of encodings, there are only a few that are of interest to us because they are constructed in a systematic manner that facilitates such computer operations as those of arithmetic. The others are mostly random encoding patterns that have little to recommend themselves for computer application.

The most common of the binary-coded-decimal-digit forms is that in which successive digits, from right to left, are weighted by successive powers of 2, the decimal digit being equal to the sum of the product of each weight times the corresponding binary digit. This is the code that we used in the radix conversions of Section 2.2. It is sometimes called the *8421* code, from the column weights. It is also often referred to as "the BCDD" form, although, as we have pointed out, there are many possible

TABLE 2.5 THE 16 POSSIBLE
CONFIGURATIONS OF FOUR
BINARY DIGITS

0 0 0 0
0 0 0 1
0 0 1 0
0 0 1 1
0 1 0 0
0 1 0 1
0 1 1 0
0 1 1 1
1 0 0 0
1 0 0 1
1 0 1 0
1 0 1 1
1 1 0 0
1 1 0 1
1 1 1 0
1 1 1 1

TABLE 2.6 ALLOWABLE
SETS OF POSITIVE WEIGHTS
FOR 4-BIT ENCODINGS OF THE
DECIMAL DIGITS

3	3	2	1
4	2	2	1
4	3	1	1
5	2	1	1
4	3	2	1
4	4	2	1
5	2	2	1
5	3	1	1
5	3	2	1
5	4	2	1
6	2	2	1
6	3	1	1
6	3	2	1
6	4	2	1
7	3	2	1
7	4	2	1
8	4	2	1

encodings of the decimal digits in binary form, and all of them may quite correctly be called BCDD forms.

A prominent advantage of the 8421 code is that it is an *additive* code; that is, the sum of the codes for two digits is the code for the sum of the two digits (subject to some adjustment if that sum exceeds 9, a matter that we examine in a later section).

Other weighted 4-bit codes are also possible. The choice of weights is not entirely arbitrary, however, and it is easy to see that the sum of the weights (if we assume only positive weights) must be at least as great as 9 so that the largest digit can be represented. Furthermore, one of the weights must be 1 (to represent decimal 1), and another must be either 1 or 2 so that decimal 2 can be represented. It follows that the sum of the four weights cannot exceed 15. The set of all possible positive weights for 4-bit encodings is shown in Table 2.6. Among these sets of weights, only the 8421 set gives unique encodings for each of the 10 decimal digits. For example, the 7421 code has two ways of representing the digit 7 as a sum of various of its weights, and the 3321 code has three ways of representing 3, two ways for 4, two ways for 5, and three ways for 6.

An interesting extension of the weightings is possible if we allow negative weights as well as positive. In this case, it is not necessary that any weight be 1, provided that we include positive and negative weights whose sum is 1. Positive weights adding "up" to at least 9 must be present, and the sum of the positive weights must not exceed 15. (If it does exceed 15, some of the 10 decimal digits cannot be represented.) Seventy-one 4-bit weighted codes having one or two negative weights are known to exist, and a complete list has been published [6].

REFERENCES

1. Couleur, J. F., BIDEC—a binary-to-decimal or decimal-to-binary converter, *IRE Trans. El. Comp.*, vol. EC-7, no. 4, Dec. 1958, pp. 313–316.
2. Lanning, Walter C., Automata for direct radix conversion, in *Computers and Electrical Engineering*, vol. 1, Pergamon Press, Elmsford, N.Y., 1973, p. 281.
3. Nicoud, J.-D., Iterative arrays for radix conversion, *IEEE Trans. Comp.*, vol. C-20, no. 12, Dec. 1971, pp. 1479–1489.
4. Bredeson, J. G., A cellular array for integer and fractional BCD–binary conversion, *Comp. Design*, May 1974, pp. 104, 106–108.
5. Benedek, M., Developing large binary to BCD conversion structures, *IEEE Trans. Comp.*, vol. C-26, no. 7, July 1977, pp. 688–700.
6. Weeg, G. P., Uniqueness of weighted code representations, *IRE Trans. El. Comp.*, vol. EC-9, no. 4, Dec. 1960, pp. 487–489.

EXERCISES

2.1. In each of the following equivalences, find the value of the radix r in which the second form of the number is expressed.
(a) $496_{10} = 1306_r$
(b) $249_{10} = 13B_r$
(c) $1248_{16} = 11110_r$
(d) $1248_{16} = 1021020_r$
(e) $228_{10} = 1403_r$

2.2. Without converting to base 10 as an intermediate step, perform the following radix conversions.
(a) $364_7 \rightarrow ($ $)_3$
(b) $755_9 \rightarrow ($ $)_5$
Check your work by converting each result back to the original base, again without converting to base 10 along the way.

2.3. (a) Given that the binary equivalent of $1/10$ is the infinitely repeating group $0.0001100110011 \ldots$, what are the binary equivalents of $2/10$? $4/10$? $8/10$?
(b) Use the successive multiplication procedure (Algorithm 2.2) to find the binary equivalents of $3/10$, $5/10$, $7/10$, and $9/10$.
(c) Verify these by binary addition of appropriate groups from part (a).

2.4. The most convenient number bases for communicating with a computer are those that are equal to an integer power of 2, since when each radix r digit in such a number is replaced by its own binary equivalent, the resulting sequence is the binary equivalent of the given number. Use this fact to convert the following base 16 numbers to binary and then from binary to octal.
(a) F13A2
(b) AB23CD

2.5. Using the procedures of Algorithms 2.7 and 2.8, show bit by bit the successive contents of the binary register and the BCDD stages when making the following conversions:

    (a) Decimal 0.6825 (expressed in BCDD form) to its binary equivalent.

    (b) Binary 0.101101110011 to three binary-coded decimal digits. (If the decimal equivalent needs more than three digits, show the truncated result; that is, use only three decimal stages.)

2.6. **(a)** Express the decimal number 275 as a base 7 number.

    (b) Now express the decimal fraction $^{275}\!/_{343}$ as a base 7 number. (It has only three digits!)

2.7. If we want to be able to express any three-decimal-digit number in base 7, with the base 7 digits each represented by a binary group, how many bits will there be in the resulting binary-coded base 7 number?

2.8. Express in octal form (i.e., base 8) the first four octal digits of the fraction $(^4\!/_{11})_{10}$

2.9. **(a)** To express a number N in octal form takes K times as many digits as are needed to express it in hexadecimal form. What is the value of K?

    (b) Fill in the blanks below.

To express numbers having:	We need to use:
3 hex digits	_____ binary digits
	_____ octal digits
	_____ decimal digits
27 binary digits	_____ hex digits
	_____ octal digits
	_____ decimal digits
10 decimal digits	_____ 8421BCD digits
	_____ binary digits
	_____ hex digits

2.10. **(a)** Express in BCDD form (8421 code) the first three decimal digits of the value of the fraction $^4\!/_7$.

    (b) Express in octal form the first three octal digits of that quantity.

    (c) Express in base 7 the first three digits of that quantity.

2.11. Suppose that we have a hardware structure for converting n-decimal-digit fractions to binary form (perhaps as in the structure of Fig. 2.2). Assume that the binary register has $\lceil n/\log_{10} 2 \rceil$ stages. Now let us suppose further that we do not have circuitry for converting decimal integers but we want to use this fraction converter. We will put our n-digit integer into the stages that would normally hold digits of the fraction and carry out the conversion in the usual way. Thus we are in effect treating our decimal integer X as if it were $X \times 10^{-n}$. To get the true result we should multiply this by 10^{+n}, but the closest we can come (in a binary register) is a shift of $\lceil n/\log_{10} 2 \rceil$ bit positions. To be specific, let us form the binary equivalent of 0.625 in our hardware, with 10 bits available in the binary register. If we then assume the binary point to be at the right (in effect, shifting it 10 bit positions), what is the absolute error between the number formed and the true binary equivalent of 0.625?

Addition Methods in Fixed-Base Number Systems

3.1 INTRODUCTION

The number systems that we use in computers commonly have such characteristics as the following:

1. They are weighted positional systems—usually.
2. They are based systems—usually.
3. They are fixed-base systems; that is, the same base is used in all positions—usually.
4. They are finite systems; that is, only a finite number of number representations are available—always.

Although in later pages we shall examine some number systems that do not have properties 1, 2, or 3, property 4 characterizes all computer number systems simply because the computer itself is a finite physical piece of equipment. This finiteness of the number system implies that there will be at least ambiguity if not also inaccuracy when the user of the computer attempts to map numbers from the infinite domain of numbers onto the finite set of computer representations.

Given that we have only a finite set of number representations available in the computer, how should we choose to map the infinite set of real numbers onto this finite set? Of course, much of the responsibility for determining this falls on the programmer, who should know something about the numbers he expects to encounter in his program, but the designer can provide means for simplifying the programmer's task. Typically, the designer must answer the following questions:

1. Should the numbers in the machine be fractions, integers, or mixed numbers?
2. Should the format be floating point or fixed point?
3. How should negative numbers be represented?
4. What number of digits should be provided for number representations?
5. What algorithms are appropriate for executing the operations of arithmetic in the number system chosen?

3.2 SIGNS AND COMPLEMENTS

Let us assume for the moment a fixed-base number system in which each number in the computer has n radix r digits. Such a group is often called a *word*, and it is clear that a word can represent any one of r^n different numbers, in combinations in the following range:

$$
\begin{array}{ccccc}
0 & 0 & 0 & \cdots & 0 & 0 \\
0 & 0 & 0 & \cdots & 0 & 1 \\
& & \vdots & & & \\
(r-1) & (r-1) & (r-1) & \cdots & (r-1) & (r-1)
\end{array}
$$

In *fixed-point* notation, the radix point is regarded as fixed at the same location in all words—at the left if the numbers are all fractions, at the right if they are all integers, and somewhere in between if they are mixed numbers. A word can alternatively represent a *floating-point* number if some of its digits represent a magnitude and the others an exponent of an implied base value, so that the word represents a number equal to

$$\text{magnitude part} \times (\text{base value})^{\text{exponent part}}$$

Whichever of these forms we employ, it is the usual practice to let half of the number representations correspond to positive numbers and half to negative numbers, as shown in Fig. 3.1 for integers. This is particularly straightforward in the case of binary numbers, since all numbers in the lower half of the set of representations will have their leftmost bit 0, and all numbers in the upper half will have their leftmost bit 1.

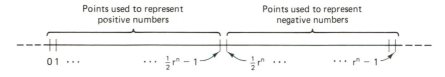

Figure 3.1 r^n points along the number line, half of them for positive numbers and half for negative.

The left bit may thus serve as a sign bit $(0 = +, 1 = -)$. However, the negative numbers can be represented in several ways:

1. The negative numbers from $-(\frac{1}{2})r^n$ through -1 are translated (i.e., moved) onto the set of representations $+(\frac{1}{2})r^n$ through $r^n - 1$. This is equivalent to adding r^n to each negative number.
2. The negative numbers from $- [(\frac{1}{2})r^n - 1]$ through -1 are translated onto the set of representations $(\frac{1}{2})r^n$ through $r^n - 2$. This is equivalent to adding $r^n - 1$ to each negative number.
3. The negative numbers from $- [(\frac{1}{2})r^n - 1]$ through -0 are reflected and translated, so that -0 maps onto $+(\frac{1}{2})r^n$ and $- [(\frac{1}{2})r^n - 1]$ maps onto $+ (r^n - 1)$.

These methods are, respectively, the *radix complement method*, the *diminished radix complement method*, and the *sign-and-magnitude method* (usually called simply "sign-magnitude"). Their relationship is shown in Fig. 3.2. The magnitude $| X |$ of an integer X is related to X_{RC}, its radix complement, by

$$X_{RC} = r^n - | X |$$

Similarly,

$$X_{DRC} = (r^n - 1) - | X |$$

Note that r^n is equal to the digit 1 followed by n 0's, and $(r^n - 1)$ is equal to a string of n digits, each equal to $(r - 1)$. When the radix $r = 2$, these are known as the *2's-complement* and the *1's-complement* forms, respectively. Similar definitions apply to fractions:

X_{RC} = radix complement of an n-digit fraction $X = 1 - | X |$

X_{DRC} = diminished radix complement of an n-digit fraction $X = (1 - r^{-n}) - | X |$

Rules for generating the complement forms may be developed by executing the indicated subtractions:

$$X_{DRC}: r^n - 1 = \quad r - 1 \qquad r - 1 \qquad r - 1 \cdots \quad r - 1$$

$$| X | = \quad x_{n-1} \qquad x_{n-2} \qquad x_{n-3} \cdots \qquad x_0$$

$$X_{DRC} = r - 1 - x_{n-1} \quad r - 1 - x_{n-2} \qquad \qquad r - 1 - x_0$$

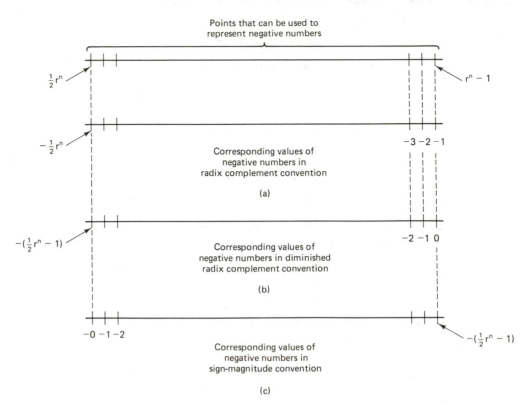

Figure 3.2 Three possible mappings of negative numbers onto the set $(1/2)r^n$ through r^{n-1}.

that is, to form X_{DRC}, subtract each digit of $|X|$ from the quantity $(r-1)$. If $r=2$, this reduces to the simple rule: Change each digit of $|X|$, (1 to 0 and 0 to 1).

$$
\begin{array}{llllll}
X_{\mathrm{RC}}: & r^n = 1 & 0 & 0 & \cdots & 0 \quad 0 \\
|X| = & x_{n-1} & x_{n-2} & & \cdots & x_1 \quad x_0 \\
\hline
& & (r-1-x_{j+1}) & (r-x_j) & \cdots & 0
\end{array}
$$

If x_0, x_1, ... are 0, the subtraction yields corresponding 0's. x_j is the assumed rightmost nonzero digit, and we subtract it from the radix by borrowing. For all subsequent j, we subtract each digit of $|X|$ from $(r-1)$. If $r=2$, this reduces to: Leave right-hand 0's and the first 1 unchanged, and change all digits to the left.

The rule for forming the 2's complement of a number is awkward to implement in the form in which we have stated it here, since it involves sequential scanning from the least significant position in order to detect the rightmost 1. The 1's complement is much easier to form, since it merely requires changing every bit. If the bits are held in flip-flops, then to transmit the 1's complement to the arithmetic unit, we simply gate from the $\overline{Q}$ outputs (the complement outputs) of all the flip-flops. However, the simple relation between the 2's complement and the 1's complement enables us to use the arithmetical advantages of the 2's complement while retaining the ease of forming the 1's complement. Note that

$$X_{\mathrm{RC}} = X_{\mathrm{DRC}} + 1$$

Thus we can form the 2's complement by adding a unit to the least significant bit of the 1's complement, a modification that is easily performed by supplying a carry bit at the input to the adder when a subtraction is executed.

In performing binary arithmetic, it is the usual practice to regard the leftmost bit as the sign, 0 for $+$ and 1 for $-$. For binary fractions, the sign bit lies to the immediate left of the radix point. Therefore, to ensure the correct formation of the sign digit as well, our rules for forming X_{RC} and X_{DRC} should involve complementation with respect to 2^1 as follows:

$$X_{\mathrm{RC}} = 2 - |X|$$
$$X_{\mathrm{DRC}} = (2 - 2^{-n}) - |X|$$

where X is regarded as having n digits to the right of the radix point. For binary integers having n digits (including the sign digit), we complement with respect to 2^n, that is,

$$X_{\mathrm{RC}} = 2^n - |X|$$
$$X_{\mathrm{DRC}} = (2^n - 1) - |X|$$

For mixed numbers having n digits (including the sign) to the left of the radix point and m to the right, we use

$$X_{\mathrm{RC}} = 2^n - |X|$$
$$X_{\mathrm{DRC}} = (2^n - 2^{-m}) - |X|$$

As we noted, splitting the set of representations into a lower range to represent positive numbers and an upper range to represent negative numbers has the result in the binary case that the leftmost bit is simply the sign bit. For decimal numbers, however, the leftmost digit is 0, 1, 2, 3, or 4 for positive numbers and 5, 6, 7, 8, or 9 for negative numbers and thus conveys some magnitude information as well as sign information. A similar observation holds for hexadecimal numbers or for octal numbers but in these cases the leftmost bit of the binary-coded-digit form is a true sign bit. For example, in a number system using four octal digits, the positive numbers are:

```
    4 octal digits            4 octal digits
    ⌒⌒⌒⌒                      ⌒⌒⌒⌒⌒⌒
    0 – – –                   0 0 0 – – –
    1 – – –                   0 0 1 – – –
    2 – – –      ⟹            0 1 0 – – –
    3 – – –                   0 1 1 – – –
                                  ⌣
and the negative numbers are:     ⎰ ═ Binary encoding of leftmost octal digit
                                  ⎱
    4 – – –                   1 0 0 – – –
    5 – – –                   1 0 1 – – –
    6 – – –                   1 1 0 – – –
    7 – – –                   1 1 1 – – –
```

As an example of these rules, consider the case of $n = 3$ and $r = 10$. Digit combinations are 000 through 999, with 000 through 499 representing positive numbers and 500 through 999 representing negative numbers. The negative number -234 is represented in radix complement form by $10^3 - 234 = 766$ and in diminished radix complement form by $(10^3 - 1) - 234 = 765$.

To negate a positive number, or a negative number in complement form, we subtract it from r^n or from $r^n - 1$ (for radix complement or diminished radix complement, respectively). Thus we would negate 766 by forming $10^3 - 766 = 234$, the positive number that is its radix complement.

Now consider 8-bit binary integers. The number -73 is represented in sign and magnitude form as 11001001. Its 2's complement form is 10110111, which we obtain by applying our bit-by-bit rule to the magnitude 01001001.

3.3 ADDITION AND SUBTRACTION

When we perform an arithmetic operation upon two number representations in the computer, we generate another digit group that is also one of the set of number representations allowable in the computer. That is, if D is the set of allowable representations, then an arithmetic operation is a mapping A from a pair $(X, Y) \in D$ to a result $Z \in D$. Ideally, we seek a simple algorithm that can be executed on the representations so that the representation of $X + Y$ will be the same as the sum (formed by that algorithm) of the representations of X and of Y, and similarly for the other arithmetic operations. Of course, one way of performing the arithmetic operations is simply to store all these mapping functions in the computer memory and to perform table lookup as needed. This is a completely impractical procedure except for very small sets of numbers. For example, an addition table for 32-bit numbers would have 2^{32} columns and 2^{32} rows, each entry being a 32-bit number, for a total of about 10^{21} bits, a number exceeding by many orders of magnitude the storage capacity of any on-line system ever proposed.

The fixed-base number system, however, lends itself nicely to table lookup based on subgroups of the sets of digits in the operands. The smallest such subgroup is a single digit, and it is an easy matter to build into the computer's circuitry either a table that stores the results of single-digit-pair addition or multiplication or else a circuit that forms those result digits when operands are applied. For fixed-base systems using radix r, we may define single-digit-pair addition in terms of a counting wheel, as shown in Fig. 3.3. To add $X + Y$ (where X and Y are single digits),

1. Enter the wheel at the value X.
2. Count up by the value of Y.
3. Count up further by the value of any carry received from the next-lower-order digit position.

Then:

4. The sum $Z = X + Y$ is found at the stopping point.
5. If the counting process *reached* or *passed through* zero, a carry of 1 to the next-higher-order digit is formed.

The same information may be displayed in an $r \times r$ table (in which entries below the diagonal are symmetric with those above and therefore need not appear).

To find the sum of two multidigit numbers, we execute this process in turn in each digit position, starting in the least significant position. Note that this procedure is essentially the same as addition by repeated determination of the successor, so that to add $X + Y$, we repeatedly replace X by the successor of X and Y by the predecessor of Y until that predecessor is 0.

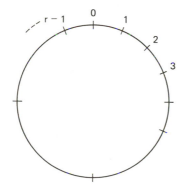

Figure 3.3 Single-digit radix r counting wheel.

Similarly, we define subtraction of $X - Y$ by:

1. Enter the wheel at the value X.
2. Count down by the value of Y.
3. Count down further by the value of any borrow occurring in the next-lower-order digit position.

Then:

4. The difference $X - Y$ is found at the stopping point.
5. If the counting process *passed through* zero (not merely stopping there), then a borrow of 1 from the next-higher-order digit is formed.

These definitions of addition and subtraction show that when negative values are formed as the result of an arithmetic operation ($+$ or $-$), they occur in the radix complement form. For example, consider three decimal "wheels" storing the number 146, as in Fig. 3.4. If we subtract 275 by the counting-down process, starting in the least significant digit, we form 871, which is the radix complement form of the number -129. In this sense, radix complements may be said to be the "natural" form for the representation of negative numbers in our fixed-base system. We note, of course, that the usual pencil-and-paper method of subtracting a number from a smaller one is instead to subtract the smaller from the larger, affixing a minus sign to the result, but the computer cannot determine which is the larger without executing a subtraction. Thus, although the signed-magnitude form is the usual one for human arithmeticians, the radix complement form is the usual one for computers.

In both the radix complement and the diminished radix complement forms, there is one digit combination that requires special attention. Consider a base 16 counter wheel, with digit values expressed in binary form, as in Fig. 3.5. By our definitions of the complement forms, we see that the diminished radix complement form has a "minus zero," represented by 1111. Half of the other values represent numbers greater than zero and half represent numbers less than zero. Arithmetic with DRC

Figure 3.4   Three decimal counting wheels, set to 146.

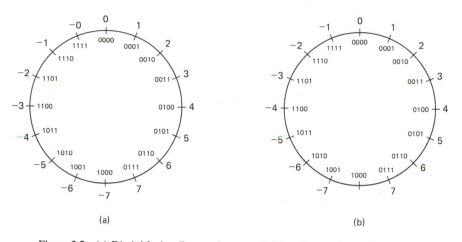

Figure 3.5   (a) Diminished radix complement and (b) radix complement forms on a base 16 counting wheel.

numbers must recognize this extra zero. If addition is done by counting, then whenever counting up reaches 1111, another 1 should be counted to form 0000, and whenever counting down passes through 0000 (not merely *to* 0000), another count down by 1 should be executed. If, on the other hand, addition is being done by the rules of the next section, we sometimes need to add a unit correction in the least significant digit. The signal to make this addition of a unit in the least significant position is the occurrence of a carry out of the most significant position. This situation is usually called an *end-around carry*, although that name is somewhat misleading, since the carry out of the most significant position is really not of the same weight as a unit in the least significant position. The situation occurs whenever we add two negative numbers or a negative number and a positive number of greater magnitude. In the first of these cases, we form

$$[(2^n - 1) - |X|] + [(2^n - 1) - |Y|] = 2 \times (2^n - 1) - (|X| + |Y|)$$

This group requires the subtraction of $(2^n - 1)$ to convert it to the DRC of $(|X| + |Y|)$, and we may regard the carry formed by 2^{n+1} as constituting the signal to make the correction. Of course, the subtraction of $(2^n - 1)$ is equivalent to the addition of 1.

In the second case we form

$$[(2^n - 1) - |X|] + Y = (2^n - 1) + (Y - |X|)$$

If $Y > |X|$, the result should be a positive value. Since 2^n itself cannot be represented in n bits, the n-bit group that the addition produces is simply $Y - |X| - 1$, and an adjustment of $+1$ is needed in the least significant bit position. The following examples are illustrative.

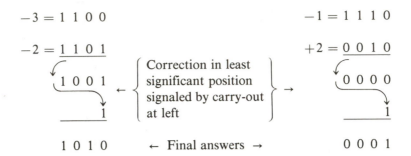

In DRC notation, the only representation of 0 that will be generated by addition or subtraction is the all-1's [or all-$(r-1)$'s] form, which is "minus zero."* That is, if we add X and its DRC, which should sum to zero, we form

$$X + [(r^n - 1) - |X|] = r^n - 1$$

The "plus zero" form, consisting of all zeros, would never be present in the computer unless it was read in as input data. Because of the inconvenience of handling two different forms of zero, the DRC form is seldom used by computer designers.

In the radix complement form, it is the group 1000 that requires our attention. If we elect to have as many number representations less than zero as greater, the 1000 group is not a member of the set of valid representations, and we can arrange for our computer circuitry to treat this group as an overflow. However, there is no necessity for having our number representations symmetrical about zero, and some designers prefer to let $100 \cdots 0$ represent a negative number of largest magnitude (i.e., of magnitude 1 unit greater than the group $0111 \cdots 1$).

If negative numbers are always represented in their radix complement forms, it is not necessary to perform both addition and subtraction in the computer. Since $X - Y = X + (-Y)$, we can subtract Y by instead adding its complement, whether Y is a positive number or a negative number in radix complement form. [Alternatively, we can make subtraction the fundamental operation, in which case $X + Y$ is formed by executing $X - (-Y)$.] Subtraction is a less convenient operation than addition, however, since it is not commutative. That is, $X - Y \neq Y - X$, and we must therefore always be careful to apply the subtrahend to the proper input point of a subtracter.

3.4 THE ADDITION PROCESS

Performing addition rapidly and with a minimum investment in equipment has been a continuing challenge to computer designers. In principle, any deterministic logic function in which the outputs are wholly determined by the present inputs can be evaluat-

---

*However, Shedletsky [1] points out an anomalous behavior that certain electrical circuits may display in DRC use.

ed in two time units—one for a group of AND gates to act after receiving the input signals and one for a group of OR gates to act after receiving signals from the AND gates. In practice, of course, the size of the gates and the number of them needed make this circuit a physical impossibility to construct, and we therefore use other methods that use less equipment but take more time. For example, to add two 32-bit numbers X and Y, where

$$X = x_{31}x_{30} \cdots x_1x_0$$

$$Y = y_{31}y_{30} \cdots y_1y_0$$

to form

$$S = S_{32}S_{31}S_{30} \cdots S_1S_0$$

we note that S_0 is a function only of x_0 and y_0 and requires only a simple logic circuit; S_1 is a function of x_1, x_0, y_1, and y_0 and requires a somewhat more complex circuit; S_j is a function of $2 \times (j + 1)$ variables; and S_{31} and S_{32} are functions of all 64 variables. To form all 33 sum digits we need 33 OR devices, some of them driven by enormously many ANDs, and some of these ANDs require up to 64 inputs. It is not feasible to gate together more than about six signals in one gate, so the direct implementation of an adder of this size in only two levels of logic is not possible.

Since the circuitry for two-level addition of large numbers is impossibly large, we resort to the common engineering step of partitioning our large unmanageable problem into smaller manageable ones whose solutions can be combined so as to solve the larger one. The price we pay is increased time for the solution, usually in the form of time delay for carry propagation. When we partition the large problem (e.g., addition of 32-bit numbers), we may go so far as to solve 32 problems each involving the addition of two 1-bit numbers. However, each such simple addition needs also information from lower-order stages, and it is this carry signal that is the cause of most of our difficulties in carrying out addition. In the following pages we examine the basic logic circuit for addition and explore various ways of handling the problem of carry propagation.

3.5 MECHANIZATION OF ADDITION

The elementary adder unit accepts a single binary digit from each of two operands and a carry digit from lower-order positions, forming a sum digit and a carry digit according to Table 3.1.

Using "+" to represent OR, "·" or merely juxtaposition to indicate AND, and an overbar to indicate NOT, we may restate the table as the following equations:

$$S_j = x_j\bar{y}_j\bar{c}_j + \bar{x}_jy_j\bar{c}_j + \bar{x}_j\bar{y}_jc_j + x_jy_jc_j \tag{3.1}$$

$$c_{j+1} = x_jy_j\bar{c}_j + x_j\bar{y}_jc_j + \bar{x}_jy_jc_j + x_jy_jc_j \tag{3.2}$$

TABLE 3.1

x_j	y_j	c_j	S_j	c_{j+1}
0	0	0	0	0
0	0	1	1	0
0	1	0	1	0
0	1	1	0	1
1	0	0	1	0
1	0	1	0	1
1	1	0	0	1
1	1	1	1	1

Applying the customary rules of Boolean algebra, we write

$$S_j = (x_j\bar{y}_j + \bar{x}_j y_j)\bar{c}_j + (x_j y_j + \bar{x}_j\bar{y}_j)c_j$$

Now let $s_j = x_j\bar{y}_j + \bar{x}_j y_j = x_j \oplus y_j$, which gives $\bar{s}_j = x_j y_j \oplus \bar{x}_j\bar{y}_j = \overline{x_j \oplus y_j}$. Then

$$S_j = s_j\bar{c}_j + \bar{s}_j c_j$$

$$S_j = s_j \oplus c_j \tag{3.3}$$

The operator "$\oplus$" denotes the exclusive-OR function. We also have

$$c_{j+1} = s_j c_j + x_j y_j \tag{3.4}$$

and we may readily show a logic diagram implementing (3.3) and (3.4) as in Fig. 3.6. Because it has three inputs x_j, y_j, and c_j, this circuit is often called a *three-input adder*. The identical subcircuits consisting of an AND and an exclusive-OR are *two-input adders*. They are also often called *half-adders*, and the complete circuit, together with its output OR, is known as a *full-adder*. (It is interesting to note that this circuit is totally symmetrical, that is, the labels of the three inputs may be interchanged in

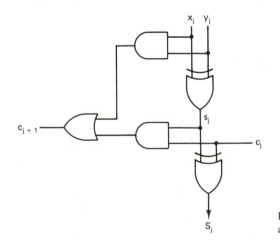

Figure 3.6 Logic structure for a full-adder stage.

TABLE 3.2

m_j	s_j	b_j	d_j	b_{j+1}
0	0	0	0	0
0	0	1	1	1
0	1	0	1	1
0	1	1	0	1
1	0	0	1	0
1	0	1	0	0
1	1	0	0	0
1	1	1	1	1

any of the six possible ways and the circuit will perform the same function. This follows from the fact that the functions S_j and c_{j+1} do not depend on what particular values the inputs have but on how many of them are 1. That is, $S_j = 1$ if and only if an odd number of the three inputs are 1, and $c_{j+1} = 1$ if and only if at least two of the inputs are 1.)

Although subtracting circuits are less commonly used than adders, it is interesting to note the close similarity of binary addition and subtraction. We let $m_j = j$th minuend bit, $s_j = j$th subtrahend bit, $b_j = $ the borrow from the jth position, and $d_j = $ the jth bit of the difference. Then the elementary three-input subtracter is defined by Table 3.2.

The difference d_j is exactly the same function of the arguments m_j, s_j, and b_j as the sum digit in an adder is of its three arguments. The borrow digit b_{j+1} is the same function of $\overline{m}_j$, s_j, and b_j as the sum digit is of x_j, y_j, and c_j. Thus it is no surprise to find that the subtracter circuit closely resembles the adder circuit. Figure 3.7 shows a three-input subtracter made of two *half-subtracters*, each consisting of an exclusive-

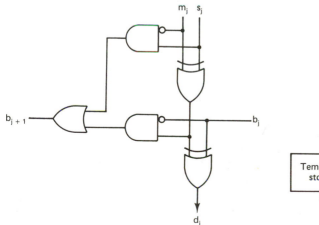

Figure 3.7   Logic structure for a three-input subtracter.

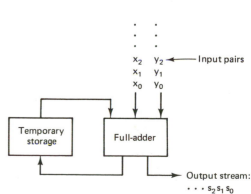

Figure 3.8   Use of a full-adder in the serial mode.

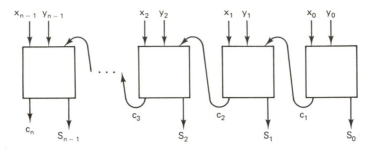

Figure 3.9 Use of n full-adders in the parallel mode.

OR element, an AND element, and an inverting element (the "bubble" at the AND inputs).

The simplest way of all to use the elementary three-input adder unit is to use it over and over to add successive digit pairs from the two operands, with each carry being stored and then applied as an input at the same time as the next input pair, as indicated in Fig. 3.8. The temporary storage is usually a flip-flop. This simple circuit requires a minimum amount of electronic logic but a maximum amount of time because of its wholly serial mode of operation.

Most applications require faster addition than is possible in the serial mode, and it is common practice to connect a chain of full adders as shown in Fig. 3.9, to accept in parallel all the digits of each of the two operands. However, in spite of having n copies of the circuit to accept at the same time n pairs of input digits, this circuit is not n times as fast. The electrical circuits of the adder stages have a time delay between their inputs and their outputs, and thus the carry digits c_j are formed sequentially. Consequently, the sum digits S_j do not take on their final correct values until all carry digits have propagated sequentially as far as they will go.

3.6 COPING WITH CARRIES

The essentially serial nature of carry propagation is the most difficult problem in speeding up addition, and a wide variety of ways have been devised for coping with it. The various techniques may be classified as follows:

1. *"Grin and bear it"*: Instead of attempting to speed up the carries, we accept them as they are, and design the adder so that its output is not utilized until the longest possible carry sequence has had time to occur, whether or not such a full-length carry sequence actually does occur.

2. *Carry-completion sensing*: With auxiliary circuitry, it is possible to sense when all carry action has been completed. At that time, the adder outputs have assumed their final correct values and can be used.

3. *Carry look-ahead*: Auxiliary circuitry can be provided to "bypass" carries around stages (or groups of stages) that would propagate a carry. Within limits, this bypassing can be simultaneous rather than sequential.

4. *Conditional-sum adder*: Like the carry look-ahead method, conditional-sum addition is an O(log n) process, n being the operand length. On successive steps it forms 1, 2, 4, 8, . . . , 2^k correct digits, until the addition is complete.

5. *Carry minimization*: Groups of k binary digits can be treated as radix 2^k numbers, and arithmetic can be carried out in this higher radix, either by direct implementation of logic circuitry or by table lookup.

6. *Carry avoidance*: Instead of using a fixed-base system of number representation, it is possible to use a nonbased system such as the residue system, which is a number representation system in which arithmetic operations are performed without any carry. Discussion of this technique is contained in Section 7.6.

7. *"Cost averaging"*: When a stream of operands are to be added, it is possible to postpone the final carry propagation until after the last operand has been entered. The time required for this deferred carry assimilation can be regarded as charged equally against all of the operands.

8. *Pipelining*: This is another form of cost averaging, suitable either for pairs of operands or for a stream of operands.

3.6.1 Characteristics of the Carry

Before we explore these methods in detail, let us examine the characteristics of the carry.

A carry will be generated at any digit position j at which

$$x_j + y_j > r - 1$$

Similarly, a carry that is received at any digit position j will cause a carry to be propagated into position $j + 1$ if

$$x_j + y_j = r - 1$$

For radix $r = 10$, Table 3.3 shows the conditions under which there will be no carry ($\times$), a carry propagated if one is received (P), and a carry generated (G). For the binary case ($r = 2$) the table reduces to merely:

or equivalently:

x_j	y_j	Carry action
0	0	No carry from this stage is possible
0	1 ⎫	
1	0 ⎭	A carry will be propagated from this stage if one is received
1	1	The stage will generate a carry

These three actions are mutually exclusive—only one of them can occur at each digit position. Thus adder stages for which both digits are 1 will at once start carry sequences. Carries propagating through stages in which $x_j \neq y_j$ will stop when they arrive at stages in which $x_j = y_j$. The following example shows carry sequences of lengths 1, 3, 1, 6, 1, 2, and 2, all starting at the same time:

$$1 \quad 3 \quad\quad 1 \quad\quad 6 \quad\quad\quad 1 \quad 2 \quad 2 \quad \text{carries}$$

$$x = 0\ 1\ 1\ 0\ 1\ 0\ 1\ 1\ 0\ 1\ 0\ 1\ 1\ 0\ 1\ 0\ 1\ 1\ 0\ 1$$

$$y = 0\ 1\ 0\ 1\ 1\ 0\ 1\ 0\ 1\ 0\ 1\ 0\ 1\ 0\ 1\ 0\ 0\ 1\ 1\ 1$$

The reader should be sure to understand that in this example *seven* different carries start simultaneously. The carry of length 1 at the left end and the carry of length 3 adjacent to it are *not* a single carry of length 4 but rather two concurrent carries. Clearly, the adder stages do not form their final correct outputs until the longest chain of carries has been completed, and in the worst case this chain of carries could be as long as the entire digit sequence. The timing of some early computers (and of present-day 8-bit microprocessors) was adjusted so that the adder outputs were not used until this longest-possible carry chain had had time to propagate, whether or not it was actually present (the "grin and bear it" method). Burks, Goldstine, and von Neumann, in their classic paper defining the structure of the general-purpose stored program

TABLE 3.3

x_j \ y_j	0	1	2	3	4	5	6	7	8	9
0	×	×	×	×	×	×	×	×	×	P
1	×	×	×	×	×	×	×	×	P	G
2	×	×	×	×	×	×	×	P	G	G
3	×	×	×	×	×	×	P	G	G	G
4	×	×	×	×	×	P	G	G	G	G
5	×	×	×	×	P	G	G	G	G	G
6	×	×	×	P	G	G	G	G	G	G
7	×	×	P	G	G	G	G	G	G	G
8	×	P	G	G	G	G	G	G	G	G
9	P	G	G	G	G	G	G	G	G	G

computer [2], analyzed the problem of carry propagation and showed that the average length of the longest carry chain to be expected in adding two n-bit numbers is of the order of $\log_2 n$. Thus they reasoned that circuitry that would detect the end of any carry action would allow the adder output to be used much sooner. Such carry-completion sensing circuits were indeed built and behaved as predicted [3]. They have found application primarily in asynchronous computers. Most present-day computers are synchronous, in the sense that flip-flops and gates are controlled by timing pulses ("clock" pulses) distributed throughout the system from a master clock pulse generator, and these pulses, rather than completion of individual circuit actions, control the next action of each circuit.

3.6.2 Carry Look-Ahead

Instead of sensing the completion of carries, it has been found practical to accelerate the propagation of carries through a method known as *carry look-ahead*.

The principle of carry look-ahead is that instead of propagating carries through the adder stages sequentially, we provide supplemental logic circuitry that forms a carry signal into each adder stage if its predecessor generated one, or if its predecessor propagated a carry generated two stages previously, or in general if its j immediate predecessors propagated a carry formed in its $(j - 1)$th predecessor. All of these actions go on simultaneously, and carry propagation thus becomes concurrent instead of sequential. However, in practice the number of stages over which this look-ahead can be applied is limited by the complexity of the gating structure, since more significant stages require successively more logic. This may be seen in the four-stage carry look-ahead circuit shown in Fig. 3.10, from which the following gate count is derived:

Stage	Number of inputs to the OR gate	Number and type of AND gates
0	—	—
1	2	One two-input
2	3	One two-input and one three-input
3	4	One two-input and one three-input and one four-input

As always, when true simultaneity cannot be achieved with reasonable amounts of equipment, we accept a slower circuit that can be realized. In carry look-ahead circuits, it is common to construct four-stage look-ahead units, each of which forms the carries to the individual stages, but which also forms a Generate output and a Propagate output to indicate the total action of the four stages. That is, if the four stages generate a carry to the fifth stage (the low-order element of the next group of four), then G = 1, and if the four stages would propagate to the fifth a carry received at the input, then P = 1. (Just as in the single-stage adder, these two conditions are mutually

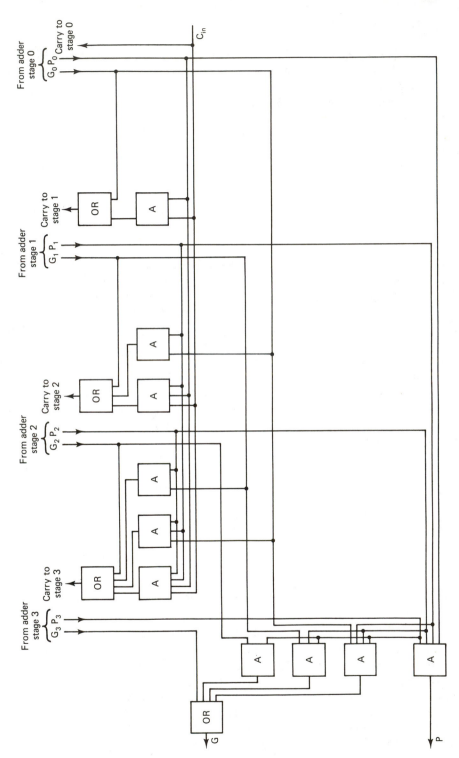

Figure 3.10 Four-stage carry look-ahead generator.

50

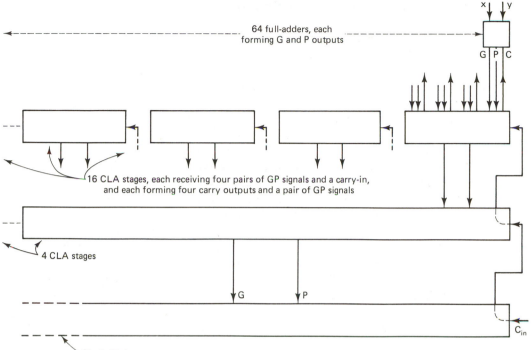

Figure 3.11 Block diagram showing use of three levels of carry look-ahead for adding 64-bit numbers.

exclusive.) These signals can then constitute inputs to a second level of look-ahead circuits. The diagram of Fig. 3.11 shows three levels of look-ahead applied to 64 adder stages. (Two levels would suffice for 16 stages, and one level for four stages.)

If we assume that each block of logic in the carry look-ahead structure can perform its functions in two gate delay times (e.g., a level of AND followed by a level of OR), then the total time required for various word lengths is:

n	T
4	$6\,\Delta t$
16	$10\,\Delta t$
64	$14\,\Delta t$

where Δt is the gate-delay time. The total consists of $2\,\Delta t$ to form the inputs to the first level of look-ahead, $2\,\Delta t$ to form the inputs to each succeeding level, $2\,\Delta t$ to propagate through the last level, $2\,\Delta t$ to propagate up through each level, and $2\,\Delta t$ to produce the final outputs. Thus the total time required is O(log n), which is a substantial improvement over the O(n) time required for ripple-carry propagation.

3.6.3 Conditional-Sum Addition [4]

A fast addition method that is closely related to the carry look-ahead techniques is the *conditional-sum adder*. In each bit position we form two tentative sum bits, one of which assumes C_{in} to that position is 0 and the other of which is based on $C_{in} = 1$. For each sum bit we also form the corresponding C_{out}. Since C_0, the carry into the rightmost position is known, we can choose the correct S_0 in this first step.

In the second time interval, the conditional sum bits from step 1 are grouped in pairs, so that for each two bit positions we form two conditional sums, each 2 bits long and each with its corresponding C_{out}. The carry from position 0 to position 1 was determined in step 1, so in the second step the correct S_1 and C_2 are determined.

In each succeeding time interval, pairs of groups are grouped, and carries into successively longer groups are formed. Thus the addition of two numbers of length 2^m requires $(m + 1)$ steps.

Here is an example in which we add two 8-bit numbers in four time intervals (Table 3.4). Corresponding to the given summands, we form at T_1 conditional sum and carry bits in each position for each assumed C_{in} at that position. These groups of digits are labeled $S_k^{C_{in}}$ and $C_k^{C_{in}}$ in the table, so that line 3, for example, shows all the sum bits formed at T_1 if corresponding C_{in} bits are 1. In our example, we will assume that C_0, the carry into the low-order position, is zero, so that at T_1 we form one correct sum digit, shown in the box at [a]. At the same time, of course, we find that the carry out of that position is also zero.

Simultaneously, conditional sum and carry bits are being formed in all positions. Consider, for instance, columns d_5 and d_4. Line 2, column d_4, shows a carry of 1, so lines 5 and 6 of column d_5 use the entries from $C_{in} = 1$ from lines 3 and 4 of column d_5. Line 2, column d_6, shows a carry of 0, so lines 5 and 6 of column d_7 use the entries from $C_{in} = 0$ from lines 1 and 2 of column d_7. We proceed in this way to form pairs of sum digits in time step T_2, each pair corresponding to a C_{in} of 0 or 1 to that pair, and each pair having a corresponding carry out. Two correct sum digits, shown in the box at [b], are formed in step T_2.

At T_3, we form conditional sum groups of length 4, and the box at [c] now contains four correct sum digits, with a carry of 1 out of the group (line 10, column d_3). This carry of 1 indicates that the correct group to choose in T_4 is the group in line 11, so at T_4 we obtain the final correct 8-bit sum, shown in the box at [d].

3.6.4 Carry Minimization by Higher Radix

The larger the number of bits for which we can simultaneously compute a sum and a carry, the shorter will be the length of the longest carry sequence and the resulting carry propagation delay. For instance, in adding two 32-bit numbers, the worst-case carry sequence is across 32 digits. If we recode the numbers as 16 base 4 numbers, we reduce the carry sequence to a maximum length of 16, and if we recode into base 2^8, doing arithmetic with 8-bit numbers, the worst-case carry sequence is only four positions. Unfortunately, the logic circuitry needed for higher radix arithmetic is much

TABLE 3.4

Time step T_k			d_7	d_6	d_5	d_4	d_3	d_2	d_1	d_0	Assumed value of C_{in} to each group of $2^{(k-1)}$ bits
Two eight-bit numbers to be added			0	1	0	1	0	1	1	0	
			0	0	1	1	0	1	1	1	
T_1 ($k=1$)	1.	S_1^0	0	1	1	0	1	1	0	[1]	$C_{in}=0$
	2.	C_1^0	0	0	0	1	0	0	1	0	$C_{in}=1$
	3.	S_1^1	1	0	0	1	1	1	1	–	$C_{in}=0$
	4.	C_1^1	0	1	1	1	0	1	1	–	$C_{in}=1$
T_2 ($k=2$)	5.	S_2^0	0	1	0	0	1	0	0	1	$C_{in}=0$
	6.	C_2^0	0		1		0		1		$C_{in}=1$
	7.	S_2^1	1	0	0	1	1	1	–	–	$C_{in}=0$
	8.	C_2^1	0		1		0		–		$C_{in}=1$
T_3 ($k=3$)	9.	S_3^0	1	0	0	0	1	1	0	1	$C_{in}=0$
	10.	C_3^0	0				0				$C_{in}=1$
	11.	S_3^1	1	0	0	1	–	–	–	–	
	12.	C_3^1	0				–				
T_4 ($k=4$)	13.		1	0	0	0	1	1	0	1	
	14.		0								

more complex than that needed for base 2 arithmetic, and it is very difficult to make it as fast as binary arithmetic circuits. An adder that accepts two 4-bit groups (each representing a base 16 digit) has nine input lines (including a carry) and five output lines (four sum digits and a carry). We cannot construct a two-level logic circuit (a group of ANDs followed by an OR) because practical electrical matters limit us to no more than about half a dozen inputs to each gate. Three-bit-position adder stages (having seven input lines and four output lines) are about the practical upper limit for implementation by two-level logic circuits.

A more attractive way of implementing higher radix arithmetic is by means of *table lookup*; that is, we provide a stored table of sum values and carries for each pair of input operands. If the input operands are k-bit numbers, such a table would have 2^k rows, 2^k columns, and 2^{2k} entries, each of which would have $(k + 1)$ bits. Thus the storage capacity needed is $(k + 1) \times 2^{2k}$ bits, or, for $k = 8$, a total of 1.125×2^{19} bits. This is too high a cost to pay in most applications, but a read-only memory for $k = 4$ would need only 1.25×2^{10} bits, which might be a quite acceptable price.

3.6.5 Deferred Carry Assimilation

Addition of two numbers X and Y can be treated as a recursive process in which we form from X and Y two new digit vectors PS^1 and PC^1, where PS^1_j, the jth digit of PS^1, is given by

$$\text{PS}^1_j = x_j \oplus y_j$$

Similarly,

$$\text{PC}^1_j = x_j \cdot y_j$$

Next we form

$$\text{PS}^2_j = \text{PS}^1_j \oplus \text{PC}^1_{(j-1)}$$

and

$$\text{PC}^2_j = \text{PS}^1_j \cdot \text{PC}^1_{(j-1)}$$

At each step in the recursion, we form a new partial sum and partial carry from the previous partial sum and the left-shifted previous partial carry. The addition is finished at the point at which the new partial carry vector is identically 0.

As an example, consider the addition of

$$X = 0\ 1\ 0\ 1\ 1\ 0\ 1$$
$$Y = 0\ 1\ 1\ 0\ 1\ 0\ 1$$

0 0 1 1 0 0 0	First partial sum	PS^1
0 1 0 0 1 0 1	First partial carry	PC^1
1 0 0 1 0 1 0	Shifted partial carry	

1 0 1 0 0 1 0	Next partial sum	PS^2
0 0 0 1 0 0 0	Next partial carry	PC^2
0 0 1 0 0 0 0	Shifted partial carry	
1 0 0 0 0 1 0	Next partial sum	PS^3
0 0 1 0 0 0 0	Next partial carry	PC^3
0 1 0 0 0 0 0	Shifted partial carry	
1 1 0 0 0 1 0	Next partial sum	PS^4
0 0 0 0 0 0 0	Next partial carry (all zeros)	PC^4

At this point, the addition is complete, since the carry digit vector is 0 in all positions.

This procedure can be mechanized either by repeated use of a group of half-adders, or by successive ranks of half-adders. In the first of these, shown in Fig. 3.12, the two registers R1 and R2 initially contain the numbers X and Y. On succeeding steps, the contents of these registers are replaced by PC^i (shifted left one position) and PS^i.

In the form shown in Fig. 3.12, the circuit is not particularly useful. It is true that we employ here only n half-adders instead of n full-adders, but the circuit action is slower than with n full-adders in a ripple-carry connection because of the time needed for the register transfer steps. However, by using three operand registers and a set of n full-adders, we obtain a structure like that of Fig. 3.13, which is extremely useful in adding a stream of operands. At each step, we add the next input vector and the partial sum and partial carry vectors from the previous operands. After the last operand has been used, the inputs to the full-adders from R2 are not needed, and gating circuits can be provided to reconnect the n full-adders in ripple-carry mode or with carry look-ahead. That is, we take time for carry propagation only once, no matter how many operands are added. By deferring the assimilation of the carries, we have

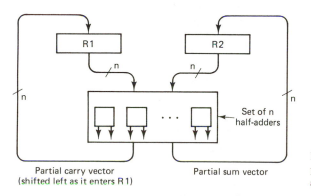

Partial carry vector
(shifted left as it enters R1)

Partial sum vector

Figure 3.12 Logic structure for adding a pair of operands using deferred carry assimilation.

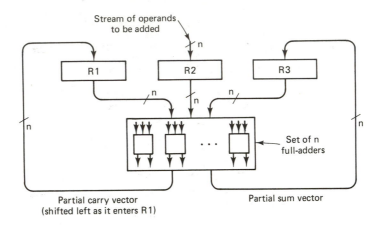

Figure 3.13 Logic structure for adding a stream of operands using deferred carry assimilation.

paid a low average price of carry assimilation time per operand. This technique, which is known both as *deferred carry assimilation* and as *carry save addition*, is a particularly useful way of speeding up the many successive additions that are done in multiplication.

It can also be useful in adding multiple operands without the intervening shift that characterizes multiplication, but in this case special care must be given to the possibility of overflow. Overflow can be handled either by assimilating the carry to detect it, or by providing a few *guard digits* beyond the most significant positions. With the guard digits, a stream of numbers of both signs might momentarily overflow the n-bit range as numbers of one sign are added and then come back into range as numbers of the other sign are encountered. Such momentary overflows cause no harm, since it is only the size of the final sum that is important. However, such overflows should be retained during the addition. To add k numbers in such a system, we should provide $\lceil \log_2 k \rceil$ guard digits at the left as temporary storage for overflow. All three registers and the set of full adders must have $n + \lceil \log_2 k \rceil$ positions, and the n-bit operands entering R2 must have their sign digits repeated in the extra $\lceil \log_2 k \rceil$ positions at the left. Suppose, for instance, that we want to add

$$
\begin{array}{l}
0.111 \\
0.111 \\
0.111 \\
0.111 \\
1.001 \\
1.001 \\
1.001 \\
1.001
\end{array}
$$

Although the sum will be 0.000, the four positive quantities will cause overflows. We therefore add three guard digits at the left, and instead perform the summation of

$$0000.111$$
$$0000.111$$
$$0000.111$$
$$0000.111$$
$$1111.001$$
$$1111.001$$
$$1111.001$$
$$1111.001$$

After adding in the four positive operands, we find that R1 contains 0011010 and R2 contains 0000010. Allowing the process to go to completion at this point without adding in the negative operands would yield 0011.100, as we would expect. Without the guard digits, this would constitute an overflow, but the guard digits allow the operation to continue, and as may be readily verified, the adding in of the negative operands will bring the sum of all eight to zero.

3.6.6 "Pipelined" Addition [5,6]

Instead of performing assimilation of carries by repeated use of the same group of half-adders, as in Fig. 3.12, we can perform the successive formation and addition of the partial sum and partial carry vectors in successive ranks of half-adders. Figure 3.14 shows such a structure for adding two 4-bit numbers. Several points may be noted:

1. Each rank of half-adders forms a final correct sum digit in the rightmost half adder.
2. There are therefore four ranks of half-adders, each rank being one unit shorter than the one above.
3. The overflow digit at the left (in this case, S_4) may be formed by an OR gate.
4. The final correct sum $S_4 S_3 S_2 S_1 S_0$ is available only after four successive half-adder time delays plus the delay of the OR gate.

A more interesting and useful form of this circuit is obtained if we introduce simple "latch" registers between the successive ranks of half-adders, as shown in Fig. 3.15. The latch registers serve as temporary storage between the rows of half-adders. A pair of operands $(X^{(1)}, Y^{(1)})$ is admitted to H_1 and the outputs of H_1 are stored in L_1. At the same time that the outputs of L_1 are applied to H_2, a second pair of operands $(X^{(2)}, Y^{(2)})$ may be admitted to H_1. Although each pair of operands must pass through all levels before its addition is complete, once the structure is full a sum is formed at the bottom each time a new pair of operands is introduced at the top.

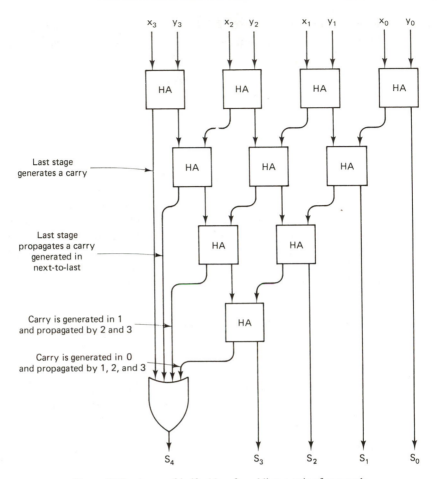

Figure 3.14   Array of half-adders for adding a pair of operands.

For obvious reasons, such a circuit is said to be a *pipeline* structure. The principle of pipelining can be used in the computer wherever a set of resources is used over and over with the output being fed back into the input a fixed number of times. Such a structure can be made into a pipeline by removing the feedback and putting k copies of the processor in sequence, each output supplying the input to the next. (k here is the number of times the original processor is used to process the initial inputs in the feedback configuration.) This is essentially what we did to convert the feedback circuit of Fig. 3.13 to the pipeline form of Fig. 3.15. Pipelining has its greatest advantage when the number of input items to be processed through the structure is substantially greater than k, the number of times the process is to be performed on each input item, since then the time required to load the structure and the time to purge it may be

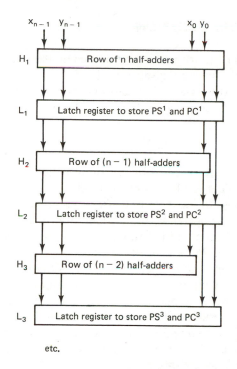

Figure 3.15 Pipelined array of half-adders for adding a stream of pairs of operands.

charged (as costs) over a larger number of elements. In the case of the pipelined 4-bit adder of Fig. 3.15, Table 3.5 shows the inputs $I_1, I_2, \ldots$ and which half-adder stage $H_1, H_2, \ldots$ they are in at each time step $T_1, T_2, \ldots$, for a total of six inputs. Note that during the loading and the purging phases, the structure is only partially occupied, but during the steady-state phase, all stages are occupied. Clearly, to minimize the fraction of time in which the stages are only partially occupied, the steady-state interval should be as long as possible.

TABLE 3.5

Stage	T_1	T_2	T_3	T_4	T_5	T_6	T_7	T_8	
				Time					
H_1	I_1	I_2	I_3	I_4	I_5	I_6			
H_2		I_1	I_2	I_3	I_4	I_5	I_6		
H_3			I_1	I_2	I_3	I_4	I_5	I_6	
H_4				I_1	I_2	I_3	I_4	I_5	I_6
		Loading phase			Steady state (full occupancy)			Purging		

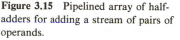

3.7 A LOGIC STRUCTURE FOR ADDITION

A common organization for the arithmetic unit of a computer employs an accumulator register AC, which is the source register for one of the operands and the destination register for the result. At the start of an addition operation, the second operand is assumed to be in register R1. As shown in Fig. 3.16, the subsequent data flow causes the operation

$$AC \leftarrow AC + R1$$

That is, the contents of AC are specified by the sum of the contents of AC and the contents of R1. The n-bit adder can be any of the mechanisms we have already examined that accepts two n-bit vectors and forms an $(n + 1)$-bit output vector. Gates (not shown in the figure) admit the operands to the adder and the result to AC. Since AC is required to receive information that is a function of its own value, some temporary storage is needed in that loop. We assume either a latch register at the adder outputs, or the use of master-slave flip-flops for AC so that the outputs of AC can be read at one set of terminals while new inputs are being applied at another.

We assume that any negative operands are represented in the 2's-complement form, and that if a negative result occurs, it too should be in 2's-complement form. r_0 and a_0, the leftmost bits of the two registers, are the sign bits. It is convenient but not necessary to assume also that the contents of R1 and AC are fractions, so that the radix-point position is immediately to the right of the sign bit. However, there is no hardware device that represents the radix point, and the hardware structure is not affected by any choice the programmer makes for the position of the radix point.

In order to determine whether the result of the operation comprised an overflow beyond register capacity, we need a record of the original sign of AC, the original sign of R1, and the final sign of AC. Since the original a_0 is replaced by the final a_0, we first make a copy of it in S. Our procedure is thus:

1. Set flip-flop S to the value of a_0.
2. Admit n-bit contents of R1 and n-bit contents of AC to the adder.
3. Ignore carry out of the adder, and admit the n-bit adder output to AC.
4. Determine overflow conditions from Table 3.6.

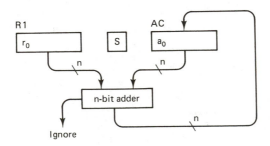

Figure 3.16 Logic structure for addition, using an accumulator register.

TABLE 3.6

S	r_0	Final a_0	Result
0	0	0	Positive; no overflow
0	0	1	Overflow occurred
0	1	0	Positive
0	1	1	Negative No overflow
1	0	0	Positive possible
1	0	1	Negative
1	1	0	Overflow occurred
1	1	1	Negative; no overflow

To understand the table, we note that addition of two positive numbers X and Y whose sum ≥ 1 will cause the final a_0 to be 1, as shown in the second line. If the operands are of opposite sign, we are forming a sum given by

$$(2 - |X|) + Y = 2 + (Y - |X|)$$

If $Y \geq |X|$, a_0 will be 0, and the quantity 2 constitutes a carry out of the adder at the left, which is ignored. But if $Y < |X|$, the result is simply the 2's complement of $Y - |X|$, in its correct form. If, however, X and Y are both negative, we are forming

$$\text{sum} = (2 - |X|) + (2 - |Y|)$$
$$\text{sum} = 4 - (|X| + |Y|)$$

If $|X| + |Y| > 1$, this sum < 3 and $a_0 = 0$, showing an overflow. If $|X| + |Y| < 1$, no overflow occurred, the sum > 3, and $a_0 = 1$. The case of $|X| + |Y| = 1$ requires our careful consideration, however. In this case, $a_0 = 1$ and all the other digits of AC are 0. If we have elected, in our choice of number representations, to have exactly as many numbers less than 0 as greater than 0, then the 1 0 0 . . . 0 case represents an overflow beyond machine capacity. To detect this condition, our overflow test must examine all digits of AC, not just a_0. If, however, we have chosen to accept the group 1 0 0 . . . 0 as representing a negative number of largest size, this case does not constitute an overflow and our overflow test need not examine all digits of AC. For this reason, it is the usual practice in machines that execute 2's-complement arithmetic to treat the group 1 0 . . . 0 as being $-2^{(n-1)}$ for integers or as -1 for fractions. With this choice, the overflow condition is indicated simply by (1) $S = r_0$, and (2) the final a_0 is different from these, that is,

$$(S = r_0) \cdot (a_0^f \neq S)$$

where a_0^f denotes the final a_0.

Although our discussion of overflow has assumed that the operands were fractions, it is equally valid for other positions of the radix point. If the radix point is taken to be k positions from the left instead of one position from the left, then each "2" in the foregoing discussion may be replaced by "2^k," without affecting the conclusions about overflow conditions.

It frequently happens that n, the length of the input vectors accepted by the adder, is much less than the length of the numbers to be added. For instance, 8 bits is a common word length in microprocessors, but many computations use 32-bit numbers. In such a case, we would partition the summands X and Y into four groups $X_3 X_2 X_1 X_0$ and $Y_3 Y_2 Y_1 Y_0$. Corresponding parts would be added in turn, with the carry from $X_j + Y_j$ being saved and added in with $X_{j+1} + Y_{j+1}$. In this situation, the carry out of the n-bit adder is no longer ignored (as it was in Fig. 3.16) but is made available to the programmer for use in adding the next pair of components $X_{j+1} + Y_{j+1}$. In the last step, when X_3 and Y_3 are added, the sign bits are included, and the final test for overflow can be made.

3.8 MULTIPLE-OPERAND ADDITION [7–10]

As we have seen, true simultaneous addition with all outputs occurring together and in two gate delay times is not possible even for only two operands, and for that case we were obliged to accept the sequential generation and propagation of carries. Some degree of sequential action is necessary in any addition method, but when we wish to add many operands, we have two choices as to the way of incurring this necessary penalty of serializing:

1. Admit the operands to the adder one after the other, with all bits of each operand transmitted in parallel.
2. Admit successive bits of each operand one after the other, with all bits in the same position in all operands admitted together.

Each of these two types of process may be further subdivided into groupings of bits and operands, so that a rather wide variety of methods result. We may visualize the array of m operands, each having n bits, as an $n \times m$ matrix:

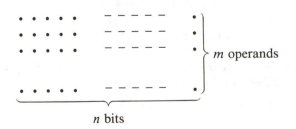

where the dots represent individual bits.

In the deferred carry assimilation method, which we have already examined, the operands are admitted sequentially, each in parallel form. This is a type 1 procedure, with no limit on m, the number of operands (except as determined by the number of guard digits). At each step, a new operand is added to the partial sum and partial carry resulting from all previously applied operands, and after the last operand is processed, the carries are allowed to go to completion, possibly in a carry look-ahead adder.

In the second type of addition, in which some (or all) of the input operands are available simultaneously, we may partition the $n \times m$ input bits into sets that can be processed separately, with the results in turn being partitioned for separate processing, until the entire $n \times m$ array has been processed.

Consider first the problem of counting the number of 1's among bits that are of the same weight, to get a count expressible in q bits. Clearly, $q = \lceil \log_2 p \rceil$. For small values of p, such a (p, q) counter can be implemented in two logic levels and will therefore have only two gate-delay times. The $(3, 2)$ counter (full-adder) is a well-known example. However, due to fan-in limitations in typical gates, it is difficult to implement a $(7, 3)$ counter in only two logic levels, and larger values of p and q are quite out of the question with present-day electronic gate technology. In any case, there will be some practical limit to p, and if we wish to count more than p bits, we must partition those bits into groups of p and use several (p, q) counters whose outputs can then be combined in a second level of logic (followed, if need be, by yet further levels of logic).

We examine the problem of counting the number of 1's among a group of 15 bits, using $(3, 2)$ counters (Table 3.7). The bits (shown by the dots) can be grouped in threes and applied to $(3, 2)$ counters, each of which can produce a bit of weight 2^0 and one of weight 2^1. In the second adder level, we combine bits of weight 2^0 as inputs to $(3, 2)$ adders and bits of weight 2^1 as inputs to other $(3, 2)$ adders.

Clearly, the limited size of our counter has forced some serializing of the counting, and in this instance five time steps were needed. The actual number of steps for any particular number m of equal-weight bits can be found as in the example, if it is needed. The number of time steps is of order of $\log m$, in general.

If more powerful counters are available, fewer time steps are needed. A $(15, 4)$ counter would count the 1's among 15 bits in only one time step, yielding the 4-bit result. If we had $(7, 3)$ counters, $(3, 2)$ counters, and $(2, 2)$ counters, we could proceed as in Table 3.8.

Consider now the problem of adding m numbers, each of n bits. One way to do this is to express the number of 1's in each column as a binary number and then to add these new smaller binary numbers, each being displaced one column from its neighbors. For example, to add six 8-bit operands, we count the 1's in each column and write the 3-bit counts with suitable displacement as in Table 3.9. Note that we have reduced the number of bits from $n \times m$ in the original array to $n \times \lceil \log_2 m \rceil$ in the second array. This process can be repeated as often as we wish, until we reach an array small enough to handle by conventional adders.

Serial addition of the m operands can be done by scanning the $n \times m$ array a

TABLE 3.7

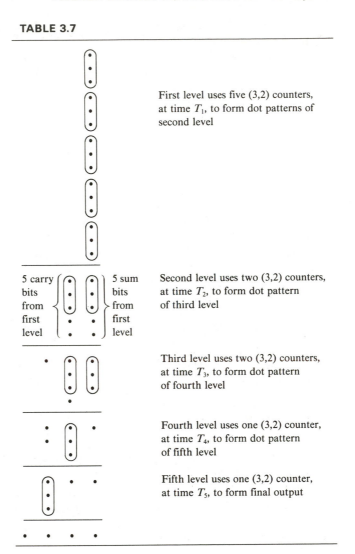

First level uses five (3,2) counters, at time T_1, to form dot patterns of second level

5 carry bits from first level

5 sum bits from first level

Second level uses two (3,2) counters, at time T_2, to form dot pattern of third level

Third level uses two (3,2) counters, at time T_3, to form dot pattern of fourth level

Fourth level uses one (3,2) counter, at time T_4, to form dot pattern of fifth level

Fifth level uses one (3,2) counter, at time T_5, to form final output

column at a time, starting from the right, and including in each column the bits of the counts of previous columns. That is, S_0, the least significant digit of the sum of the m operands is the least significant digit of K_0, the count of 1's in the d_0 column. S_1 is the least significant digit of the count of 1's in the d_1 column and the d_1 position of K_0. S_2 is the least significant digit of the count of 1's in the d_2 column and in the d_2 position of K_0 and in the d_2 position of K_1. If we assume for the moment a counter that can count a large number of inputs in one time step, the action of our process would be as shown in Table 3.10.

The structure of Table 3.10 can be mechanized by a counter augmented by a

TABLE 3.8

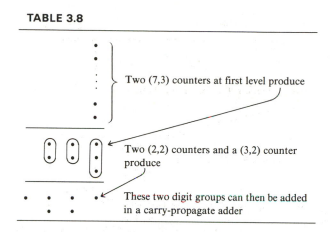

Two (7,3) counters at first level produce

Two (2,2) counters and a (3,2) counter produce

These two digit groups can then be added in a carry-propagate adder

TABLE 3.9

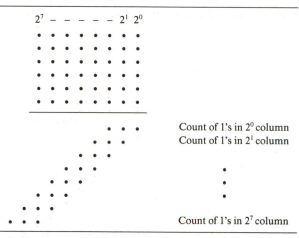

Count of 1's in 2^0 column
Count of 1's in 2^1 column

Count of 1's in 2^7 column

TABLE 3.10

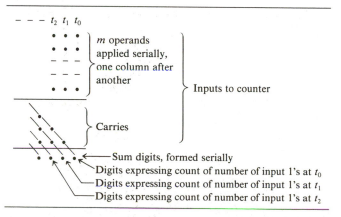

m operands applied serially, one column after another

Inputs to counter

Carries

Sum digits, formed serially
Digits expressing count of number of input 1's at t_0
Digits expressing count of number of input 1's at t_1
Digits expressing count of number of input 1's at t_2

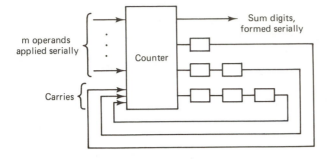

Figure 3.17 Structure for adding m operands, applied simultaneously but serially.

feedback path with appropriate delay for each carry, as in Fig. 3.17. The small rectangles represent unit time delays such as latch flip-flops.

In the likely event that a counter is not available that can accept a large number of inputs, we can synthesize it from smaller counters such as the (3, 2) counter. We group the m inputs into pairs (possibly with one left over, if m is odd) and apply these pairs to (3, 2) counters each having a delayed carry feedback. The outputs of these are similarly paired as inputs to another level of (3, 2) counters, and so on, until the counting has been completed. The structure is shown in Fig. 3.18 for $m = 8$, and is clearly adaptable to the pipelining of its functions.

The serial addition of $m = 2^k$ operands can be done with the use of $1 + 2 + \cdots + 2^{k-1} = 2^k - 1$ counters of (3, 2) type. The time required will be n plus the number of levels in the tree, which will be $\lceil k \rceil$. Although this method achieves economical use of counters, it may be too slow for some applications. The fastest way to add the m operands, using only (3, 2) counters and (2, 2) counters, is to use as many counters as we can on the $n \times m$ array, and then to use another set on the outputs of these, and so on. For example, to add seven 8-bit numbers, we could proceed as shown in Table

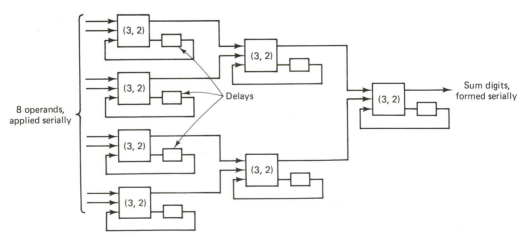

Figure 3.18 Eight-input counter with serial output.

TABLE 3.11

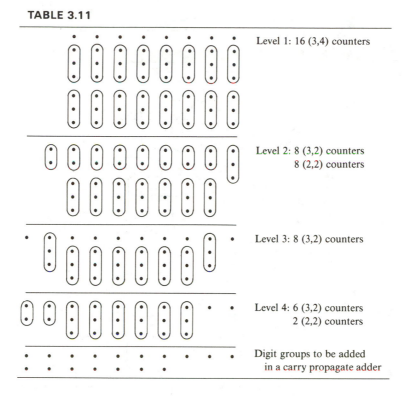

Level 1: 16 (3,4) counters

Level 2: 8 (3,2) counters
        8 (2,2) counters

Level 3: 8 (3,2) counters

Level 4: 6 (3,2) counters
        2 (2,2) counters

Digit groups to be added
    in a carry propagate adder

3.11. Four levels of counters are needed to form finally two bit patterns that can be added in a conventional carry-propagate adder. Although this is a faster method than serial addition, we pay the heavy price of using 38 (3, 2) counters and 10 (2, 2) counters.

A variant of this technique is due to Lim [11], who uses groups of (3, 2) counters to construct $(p, 2)$ counters (p an odd number) that have two outputs, a partial sum bit and a partial carry bit, and that also have four carry outputs that do not propagate beyond the next stage. The circuits of the (5, 2) counter and the (7, 2) counter are shown in Fig. 3.19. Note that carries go only to the next stage and do not affect the next stage's carry to subsequent stages. Note also that the output bit pairs are formed after three full-adder delays in the (5, 2) counter and after four full-adder delays in the (7, 2) counter. Indeed, if we add seven 8-bit numbers using these (7, 2) counters, we use 35 (3, 2) counters and four adder times, while in Table 3.11 we used 38 (3, 2) counters and 10 (2, 2) counters, with the same time required. Thus this structure appears somewhat more economical. It is an easy matter to extend the circuit pattern of Fig. 3.19 to cases of the (9, 2) counter, the (11, 2) counter, and the general $(2k + 1, 2)$ counter, and the reader is urged to carry this through to confirm his understanding of the circuit.

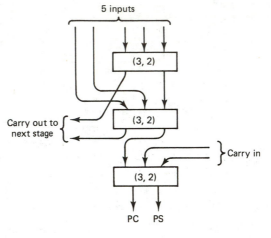

(a)

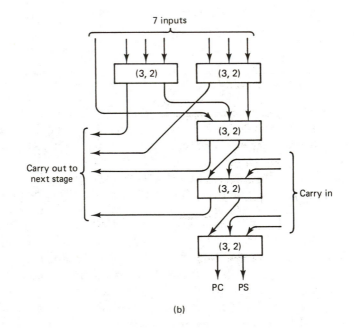

(b)

Figure 3.19   (a) A (5, 2) counter and
(b) a (7, 2) counter, using (3, 2) counters.

REFERENCES

1. Shedletsky, J. J., Comment on the sequential and indeterminate behavior of an end-around-carry adder, *IEEE Trans. Comp.*, vol. C-26, no. 3, Mar. 1977, pp. 271–272.

2. Burks, A. W., H. H. Goldstine, and J. von Neumann, Preliminary discussion of the logical design of an electronic computing instrument, Institute for Advanced Study, Princeton, N. J., 1946 (reprinted in C. G. Bell and A. Newell, *Computer Structures: Readings and Examples*, McGraw-Hill, New York, 1971).

3. Gilchrist, B., J. H. Pomerene, and S. Y. Wong, Fast carry logic for digital computers, *IRE Trans. El. Comp.*, vol. EC-4, no. 4, Dec. 1955, pp. 133–136.

4. Sklansky, J., Conditional-sum addition logic, *IRE Trans. El. Comp.*, vol. EC-9, no. 2, June 1960, pp. 213–226.

5. Hallin, T. G., and M. J. Flynn, Pipelining of arithmetic functions, *IEEE Trans. Comp.*, vol. C-21, no. 8, Aug. 1972, pp. 880–886.

6. Chen, T. C., Overlap and pipeline processing, in *Introduction to Computer Architecture*, ed. H. S. Stone, Science Research Associates, Chicago, 1975, Chap. 9.

7. Singh, S., and R. Waxman, Multiple operand addition and multiplication, *IEEE Trans. Comp.*, vol. C-22, no. 2, Feb. 1973, pp. 113–120.

8. Agrawal, D. P., and T. R. N. Rao, On multiple operand addition of signed binary numbers, *IEEE Trans. Comp.*, vol. C-27, no. 11, Nov. 1978, pp. 1068–1070.

9. Dadda, L., Composite parallel counters, *IEEE Trans. Comp.*, vol. C-29, no. 10, Oct. 1980, pp. 942–946.

10. Dormido, S., and M. A. Canto, Synthesis of generalized parallel counters, *IEEE Trans. Comp.*, vol. C-30, no. 9, Sept. 1981, pp. 699–703.

11. Lim, R. S., High-speed multiplication and multiple summand addition, *Proc. 4th Symp. on Comp. Arith.*, Oct. 1978, IEEE Cat. no. 78CH1412-6C, pp. 144–153.

EXERCISES

3.1. **(a)** Each of the following binary number pairs is to be added in a parallel adder such as that of Fig. 3.9. In each case, determine
   (1) The number of carry sequences that start simultaneously.
   (2) The length of the longest carry sequence.

   Pair 1:   1 0 0 1 1 0 1 0 1 1 0 1 0 1 1 1
             0 1 0 0 1 1 1 1 0 1 0 0 1 0 1 1

   Pair 2:   0 1 0 0 1 0 0 1 0 1 0 1 1 0 1 1
             0 1 0 1 0 1 1 0 1 1 0 1 0 0 1 0

   Pair 3:   1 0 1 0 1 1 1 0 1 0 1 1 0 1 0 1
             0 1 0 1 1 1 1 1 0 1 1 0 0 1 0 1

(b) Instead of performing the addition using 16 base two adders, use four base 16 adders. For each of the pairs above, how many carry sequences would now be initiated, and what would be the length of the longest?

3.2. Carry look-ahead over 4-bit groups, and then over four groups of 4-bit groups, and then over four such groups, results in optimum operand lengths of 4, 16, 64, and so on. Other lengths are possible, of course, but the speed-up benefits are not so great as for these optimum values.

(a) How many 4-bit look-ahead units are needed for each of the following operand lengths?

(b) What is the total number of levels of CLA in each case?

(c) What is the total number of gate delay times for completion of an addition?

$$\text{Operand length:} \quad \begin{array}{l} 18 \text{ bits} \\ 32 \text{ bits} \\ 48 \text{ bits} \end{array}$$

3.3. Table 3.6 shows a way of expressing the overflow condition in terms of the signs of the operands and the sign of the result. In many microprocessors, the overflow condition is determined from the carry into and the carry out of the sign bit position:

$$\text{overflow} = C_{\text{out}} \oplus C_{\text{in}}$$

Show that this condition is equivalent to that given in Table 3.6.

3.4. The structure of Fig. 3.16 can also be used to subtract the contents of R1 from AC. We customarily do this by complementing every bit of R1 and then inserting a carry C_0 into the parallel adder when the addition is performed. How should Table 3.6 be modified when subtraction is performed? Assume that S again stores the original sign bit of AC and that r_0 is the complement of the original sign of R1.

3.5. The structure of Fig. 3.16 can use the flip-flop S as a "signs same" flip-flop; that is, it is set to 1 if the initial signs of R1 and AC are the same and is set to 0 if they differ. Modify Table 3.6 for the addition case and for the subtraction case when we use flip-flop S in this way. For subtraction, assume that S is set (to 1 or 0, as the case may be) *before* R1 is complemented, and that the value of r_0 in the table will be the value *after* complementing R1.

3.6. The structure of Fig. 3.16 can also be used to add numbers that are in 1's-complement form. To detect the need for end-around carry, we now use the carry-out signal from the adder, which we ignored in 2's-complement addition. The carry-out signal from the left may be connected to the C_0 input at the right. Show that with this connection made, the longest possible carry chain is still only n bit positions; that is, the end-around carry that is introduced at the right cannot propagate beyond the rightmost stage that initiated a carry. What modifications (if any) are needed in Table 3.6 for 1's-complement addition?

3.7. **(a)** Draw the circuit of a (9, 2) counter made up of (3, 2) counters as in Fig. 3.19. How many carry lines will such a counter have to the next stage? What time delay will be involved in forming the PC and PS outputs?

(b) If a group of these counters are used to add nine 8-bit numbers, the low-order stages will not use or form as many carries as the high-order stages. How many (3, 2) counters will be needed in the least significant bit position? How many in the next position?

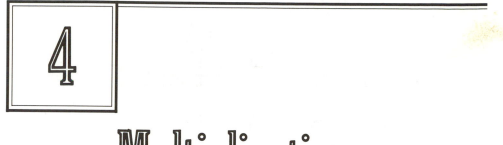

4

Multiplication

4.1 INTRODUCTION

The mapping operation that transforms a pair of number representations (X, Y) into Z, the representation of the product, is a more complex one than that defining addition. D, the set of representations, has r^n members (where n is the number of radix r digits in each number), but the product of two n-digit numbers is a $2n$-digit number that belongs to a set D' having r^{2n} members. Such a set is easily derived from the set D by providing a second n-digit group that subdivides the r^n subdivisions defined by D. Thus we seek algorithms that map a pair of n-digit representations $(x, Y) \in D$ to a $2n$-digit representation $Z \in D'$. (Note that this mapping of a pair in D to an element in D' does not exhaust D'; that is, there are many elements in D' that are not the double-length product of any pair in D. For example, there are 100 two-decimal-digit numbers, but only 37 of them are the product of one decimal digit by another.)

4.2 THE BASIC MULTIPLICATION ALGORITHM FOR MAGNITUDES

Such an algorithm is provided by a direct implementation of the pencil-and-paper method of multiplication.

Let the multiplicand $Y = y_{n-1}y_{n-2} \cdots y_1y_0$ ("ICAND") and the multiplier $X = x_{n-1}x_{n-2} \ldots x_1x_0$ ("IER"). Then form successively (from $j = 0$ to $n - 1$)

$$\text{ICAND} \times x_j$$

and add them, with each $\text{ICAND} \times x_j$ shifted left one position from the previous one. To simplify the addition process the computer immediately adds each shifted ICAND $\times x_j$ term to the previous partial sum instead of saving all these terms and adding them at the end. Formation of ICAND $\times x_j$ terms is especially simple in the binary system, since each x_j is either 0 or 1.

Since the product will have $2n$ digits, we must provide hardware facilities for formation of double-length products. It is customary to give the programmer access to

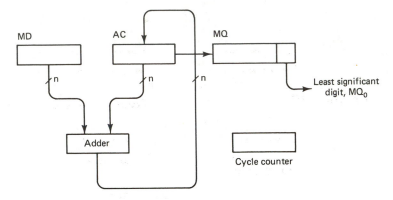

Figure 4.1 Register configuration for binary multiplication.

both parts of this product so that the decision as to their use may be his. The essential elements of the structure for performing binary multiplication are shown in Fig. 4.1. For the moment, we will assume that both operands are positive, or equivalently, that we are multiplying the magnitudes of the two operands. AC, MQ, and MD are hardware registers, each n bits long.* AC and MQ can be shifted right as a single double-length register, the rightmost bit of MQ being lost, the rightmost bit of AC entering the left end of MQ, and the previous carry out of AC entering the left end of AC. The cycle counter is used in determining the number of iterations of the loop part of the algorithm. In terms of these elements, the algorithm is:

1. Initialize.
   (a) Set cycle counter to n.
   (b) Clear AC.
   (c) Put IER into MQ register.
   (d) Put ICAND into MD register.
2. Repeat, while cycle counter $\neq 0$
   (a) $AC \leftarrow AC + ICAND \times MQ_0$.
   (b) Shift AC and MQ one position right.
   (c) Decrease cycle counter contents by 1.
3. Multiplication is complete. High-order part of double-length product is in AC and low-order part is in MQ.

It should be observed that as the multiplier is shifted to the right to bring its more significant digits successively into the MQ_0 position, the leftmost positions of MQ are thus made available to accommodate the low-order digits of the AC as they

---

* "AC," of course, denotes the accumulator. "MD" means "multiplicand-divisor," and "MQ" means "multiplier-quotient." These registers are so named because they have roles both in multiplication and in division.

TABLE 4.1

MD = 1010

AC	MQ	CC	Step
0 0 0 0	0 1 0 1	4	1
1 0 1 0	0 1 0 1	4	2(a), with $MQ_0 = 1$
0 1 0 1	0 0 1 0	4	2(b)
0 1 0 1	0 0 1 0	3	2(c)
0 1 0 1	0 0 1 0	3	2(a), with $MQ_0 = 0$
0 0 1 0	1 0 0 1	3	2(b)
0 0 1 0	1 0 0 1	2	2(c)
1 1 0 0	1 0 0 1	2	2(a), with $MQ_0 = 1$
0 1 1 0	0 1 0 0	2	2(b)
0 1 1 0	0 1 0 0	1	2(c)
0 1 1 0	0 1 0 0	1	2(a), with $MQ_0 = 0$
0 0 1 1	0 0 1 0	1	2(b)
0 0 1 1	0 0 1 0	0	2(c)

| High-order digits of the product | Low-order digits of the product | | |

are shifted out, the partial product growing in length as the remaining fragment of the multiplier shrinks. Note also that the cycle counter counts down from n to 0 instead of up from 0 to n. Either way will work, of course, but since the cycle counter is also used in other algorithms, less circuitry is involved if we always use the same test for completion (i.e., is CC = 0?), preloading the counter with the number of steps to be executed and then counting down to the common completion condition. Table 4.1 shows an example of the use of this algorithm to multiply the positive numbers 1010 and 0101. In practice, of course, decrementing and testing the cycle counter would not require a separate time step but would be done concurrently with either step 2(a) or 2(b). In the algorithms that follow, we will omit explicit reference to the cycle counter.

4.3 INCLUSION OF THE SIGN

Now consider more carefully the number of steps in this algorithm. In multiplying numbers in sign-magnitude form, we multiply the magnitudes and attach the proper sign to the result. If the two numbers are each n bits long, including their sign digits, the magnitude parts are $(n - 1)$ bits long. We may use two registers, each of length $(n - 1)$, concatenated to hold a double-length product of $2(n - 1)$ bits, or we may use two n-bit registers, with the sign bit (leftmost) set to 0 initially in each register. In either case, the correct sign for the product ($+$ if the operand signs agreed and $-$ if they differed) is stored separately and attached to the result as a final step, and in either case, the add-and-shift procedure is performed $(n - 1)$ times. The adder, of

TABLE 4.2

MD = 0101

Result sign = 1

	AC	MQ	
	0 0 0 0	0 1 1 1	Initially
	0 1 0 1	0 1 1 1	$MQ_0 = 1$, so add ICAND
	0 0 1 0	1 0 1 1	Shift (AC, MQ)
	0 1 1 1	1 0 1 1	$MQ_0 = 1$, so add ICAND
	0 0 1 1	1 1 0 1	Shift (AC, MQ)
	1 0 0 0	1 1 0 1	$MQ_0 = 1$, so add ICAND
	0 1 0 0	0 1 1 0	Shift (AC, MQ)
Answer	1 1 0 0	0 1 1 0	Insert result sign at left of AC

| Sign of result | 3 bits of high-order part | 3 bits of low-order part | Sign of multiplier |

course, accepts two $(n-1)$ bit groups. After $(n-1)$ iterations, the two n-bit registers (more commonly used than $(n-1)$ bit registers) will contain:

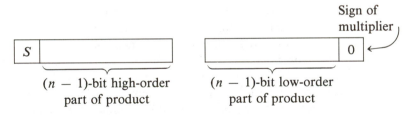

As a final step, S will then be set to the previously stored value for the sign of the result.

For example, let us multiply $+0.101$ by -0.111 (the operands being assumed to be in sign-magnitude form) (Table 4.2). Note that in executing this procedure, we do not treat the leftmost bit of the accumulator as a sign bit. That is, if a 1 should appear in the leftmost position during our procedure (as it did in step 3 of our example), it is simply shifted and replaced by a zero. If the 1 had been treated as a minus sign, as it would be in complement representation, it would have been both shifted right and copied into its original position.

4.4 TWO'S-COMPLEMENT MULTIPLICATION

Although the preceding algorithm serves to multiply the magnitudes of two numbers, it is not suitable for multiplying factors that are in two's-complement form. For example, if we multiply a positive fraction X by a negative fraction Y, we perform $X \times [2$

$-|Y|] = 2X - X \times |Y|$, but the result we want is $2 - X \times |Y|$. Similarly, for the product of a negative Y and a negative X, our algorithm gives $[2 - |Y|] \times [2 - |X|] = 4 - 2 \times (|X| + |Y|) + |X| \times |Y|$ instead of $|X| \times |Y|$. In each case an appropriate correction must be introduced to ensure the desired result.

4.4.1 Robertson's Method

Two methods of handling factors in complement form that we will examine are the methods devised by J. E. Robertson [1] and A. D. Booth [2]. Robertson's method amounts to adding the correction term bit by bit as the steps of the multiplication algorithm are executed. Consider first the multiplication of a negative ICAND Y by a positive IER X. The required correction term is $(2 - 2X)$, that is, the 2's complement of $2X$. Let X have a sign digit (which is 0, because X is positive) and $n - 1$ nonsign digits, in its fraction part. The quantity $1 - X$ is the 2's complement of the nonsign digits, and $2 - 2X$ is that quantity shifted left one position. The relative positions of the correction term and of the final high-order part and low-order part are as shown.

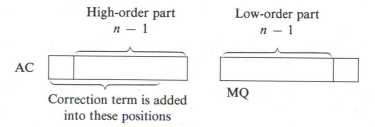

For instance, longhand execution of our magnitude multiplication algorithm, with ICAND $Y = 1.0011$ (i.e., -0.1101) and IER $X = 0.1011$ would be done as follows:

$$
\begin{array}{rl}
Y = 1.0\ 0\ 1\ 1 & \\
X = 0.1\ 0\ 1\ 1 & \\
\hline
1\ 0\ 0\ 1\ 1 & \quad 1 \times \text{ICAND} \\
1\ 0\ 0\ 1\ 1 & \quad 1 \times \text{ICAND} \\
0\ 0\ 0\ 0\ 0 & \quad 0 \times \text{ICAND} \\
1\ 0\ 0\ 1\ 1 & \quad 1 \times \text{ICAND} \\
\hline
0.1\ 1\ 0\ 1\ 0\ 0\ 0\ 1 & \quad \text{Product}
\end{array}
$$

To this result we must add the correction $2 - 2X = 0.1010$:

$$
\begin{array}{r}
+\ 0.1\ 0\ 1\ 0 \\
\hline
1.0\ 1\ 1\ 1\ 0\ 0\ 0\ 1
\end{array}
$$

which is the desired correct result, as the reader may easily verify.

The reason that the line marked "Product" is incorrect is that the steps have not taken into account the signs of the partial products being added. In each place where the negative ICAND is being added in, its sign should be extended left to the sign position of the final product. Our example is repeated below, but this time with the sign digits copied to the left.

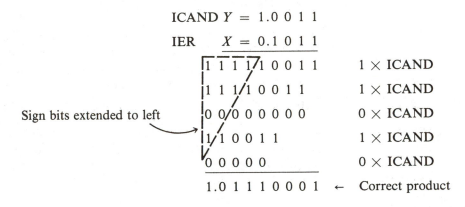

$$ICAND \; Y = 1.0\ 0\ 1\ 1$$
$$IER \qquad X = 0.1\ 0\ 1\ 1$$

Sign bits extended to left

1 1 1 1/1 0 0 1 1	$1 \times ICAND$
1 1 1/1 0 0 1 1	$1 \times ICAND$
0 0/0 0 0 0 0 0	$0 \times ICAND$
1/1 0 0 1 1	$1 \times ICAND$
0 0 0 0 0	$0 \times ICAND$

$$1.0\ 1\ 1\ 1\ 0\ 0\ 0\ 1 \quad \leftarrow \quad \text{Correct product}$$

In general, when we multiply a negative ICAND Y by a positive IER X, the accumulator contents will become negative when the first 1 of the IER has been detected, and they should remain negative through all subsequent steps. Thus, after this first IER 1, whenever AC is shifted right, its sign bit should be set to 1 after the shift, and whatever value (0 or 1) was in the sign position (as a result of adding AC + ICAND) is, of course, shifted to the right. One way to shift these 1's into the AC sign position is to provide a 1-bit register to the left of AC whose contents are copied into that position on each shift step. This flip-flop FF is initially cleared to 0 and stays at 0 until the rightmost IER 1 has been detected. It is then set to 1 and retains that value through the rest of the multiplication. Table 4.3 shows the same values of X and Y multiplied by this procedure. This method is easily extended to all sign combinations in the following way:

1. *Both factors positive*: Form $X \times Y$ in the usual way, always shifting 0 into the AC sign position regardless of whether that sign digit was 0 or 1 before the shift.
2. *IER positive and ICAND negative*: Form $X \times Y$ in the usual way, inserting 0 into the AC sign position after each shift until the first 1 in the IER is encountered, after which 1 is inserted into the AC sign position at every shift.
3. *IER negative and ICAND positive* } Form $-X$ and $-Y$ and then
4. *IER negative and ICAND negative* } use rule 1 or 2

Although in principle it is a straightforward matter to complement both operands, for cases 3 and 4, in practice it is somewhat awkward, or at least time consuming, and a

TABLE 4.3

MD

1 0 0 1 1

Sign FF	AC	MQ	
0	0 0 0 0 0	0 1 0 1 1	Initially
1	1 0 0 1 1	0 1 0 1 1	$MQ_0 = 1$, so add ICAND; set sign $= 1$
1	1 1 0 0 1	1 0 1 0 1	Shift (sign, AC, MQ)
1	0 1 1 0 0	1 0 1 0 1	$MQ_0 = 1$, so add ICAND
1	1 0 1 1 0	0 1 0 1 0	Shift (sign, AC, MQ)
1	1 0 1 1 0	0 1 0 1 0	$MQ_0 = 0$, so don't add ICAND
1	1 1 0 1 1	0 0 1 0 1	Shift (sign, AC, MQ)
1	0 1 1 1 0	0 0 1 0 1	$MQ_0 = 1$, add ICAND
1	1 0 1 1 1	0 0 0 1 0	Shift (sign, AC, MQ)

Sign of result	4 bits of high-order part	4 bits of low-order part	Sign of multiplier

better procedure is available. This consists of subtracting the ICAND after the last step when the IER sign bit is in the MQ_0 position. The validity of such a step follows directly from our definition of the 2's-complement form of the IER. If the IER is $X = x_{n-1}x_{n-2} \cdots x_1 x_0$ and its sign bit x_{n-1} is 1, then the true value of X is

$$-2^{n-1} + \sum_{0}^{n-2} x_i 2^i$$

For instance, 1 1 0 1 is the 2's-complement form of $-0\ 0\ 1\ 1$, which is $-1\ 0\ 0\ 0 + 0\ 1\ 0\ 1$.

Cases 3 and 4 now become:

3. IER negative and ICAND positive: Follow the same procedure as in case 1, but after the final step subtract the ICAND.
4. IER negative and ICAND negative: Follow the same procedure as in case 2, but after the final step subtract the ICAND.

Table 4.4 shows an example of case 4.

One matter that should be resolved by the designer and made clear to the user is the position of the $(2n - 2)$-bit product in the $2n$-bit register pair AC, MQ. We have two reasonable choices, as shown in Fig. 4.2. The arrangement at Fig. 4.2a has the advantage of symmetry—each register holds exactly half of the digits. But then the

TABLE 4.4

MD = 1011

Sign FF	AC	MQ	
0	0 0 0 0	1 1 0 1	Initially
1	1 0 1 1	1 1 0 1	$MQ_0 = 1$, so add ICAND; set sign FF = 1
1	1 1 0 1	1 1 1 0	Shift (sign, AC, MQ)
1	1 1 0 1	1 1 1 0	$MQ_0 = 0$, don't add
1	1 1 1 0	1 1 1 1	Shift (sign, AC, MQ)
1	1 0 0 1	1 1 1 1	$MQ_0 = 1$, so add ICAND
1	1 1 0 0	1 1 1 1	Shift (sign, AC, MQ)
1	0 0 0 1	1 1 1 1	$MQ_0 = 1$, so subtract ICAND

Sign of result ⟋ 3 bits of high-order part   3 bits of low-order part   ⟍ Sign of multiplier

rightmost bit of MQ must be forced to zero, since after $(n - 1)$ shifts it contains the original sign of the multiplier. We could, of course, do one more shift of MQ alone and repeat the sign of AC in the leftmost bit of MQ.

Figure 4.2b represents an alternative to both of these arrangements and has the advantage that we may now use the digit group $100 \cdots 0$ to represent a negative number of maximum magnitude. The product of that number by itself has a sign and $(2n - 1)$ nonsign bits. For example, in a 4-bit format, 1000 represents -8, and $1000 \times 1000 = 01000000$, which represents $+64$. In this one case, the two leftmost bits are 01, but in all other cases they will be 00 for positive numbers and 11 for negative numbers. The algorithms that follow will use this form for the final product.

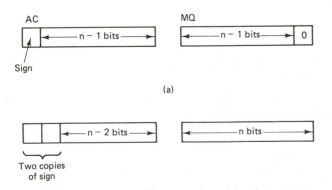

(a)

(b)

Figure 4.2 Possible positions of the $(2n - 2)$-bit product in the $2n$-bit register pair.

4.4.2 Booth's Method

In 1951, A. D. Booth [2] described another method for multiplying that accepts numbers in 2's-complement form. The procedure uses a recoding of the multiplier based on the fact that k 1's in sequence are equivalent to

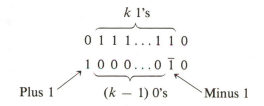

Any binary number can be recoded by this equivalence in such a way that each group of one or more adjacent 1's is replaced by a minus 1 (denoted by $\bar{1}$) at its low end, a positive 1 to left of its high end, and intervening 0's. For example:

1. (a) *Binary*:   01100111
    (b) *Recoded*: $10\bar{1}0100\bar{1}$
2. (a) *Binary*:   11100100
    (b) *Recoded*: $100\bar{1}01\bar{1}00$

Note that example 1 is the binary representation of the number 103. Its recoded form can be regarded as

$$1 \times 2^7 - 1 \times 2^5 + 1 \times 2^3 - 1 \times 2^0$$

which, of course, is also 103. A similar analysis can be made of example 2.

Consider now the recoding of an n-bit multiplier which is in 2's-complement form, with its sign bit equal to 1. Limit the recoded form to n digits, each either 0 or $+1$ or -1. The weighted sum of the positive and negative digits of this recoded number will always give a true sign-magnitude form for the original number, thus assuring that we can replace 2's-complement multipliers by their Booth recodings in executing multiplication. As examples:

1. (a) *Binary:* $1.0011010 = -0.1100110$
    (b) *Booth recoded:* $\bar{1}.010\bar{1}1\bar{1}0 = \left.\begin{array}{l} -1.0001010 \\ +0.0100100 \end{array}\right\} = -0.1100110$
2. (a) *Binary:* $1.11100110 = -0.00011010$
    (b) *Booth recoded:* $0.00\bar{1}010\bar{1}0 = \left.\begin{array}{l} -0.00100010 \\ +0.00001000 \end{array}\right\} = -0.00011010$

Because of its obvious similarities to the process of differentiation, Booth recoding is sometimes known as *differentiating recoding*.

To employ this recoding in the computer, we do not actually create a digit group having digit values of $+1$, 0, or -1. Such a representation would be a ternary form, requiring extra circuitry. Instead, the multiplier is scanned from the less significant end, and instead of performing an addition for each 1 in the IER, the circuitry is designed to perform a subtraction of the ICAND when the start of a sequence of 1's is detected and an addition when the end of that sequence is detected. Let the multiplier be $d_{n-1}d_{n-2} \cdots d_1 d_0$. The recoding is performed by examining the digits by pairs, starting with $d_0 0$ (note that we assume a 0 to lie to the right of the least significant digit), then $d_1 d_0$, then $d_2 d_1$, and so on to the left, concluding with the sign digit. The multiplication rules are:

d_{j+1}	d_j	Action in position $(j + 1)$
0	0	None
0	1	Add ICAND
1	0	Subtract ICAND
1	1	None

Following the indicated action, (AC, MQ) is shifted right one position as in our previous algorithms for multiplication.

4.4.3 Modified Booth Method

Before we show examples of this procedure, let us first make a simple and useful modification to the recoding. We observe that each nonzero digit of the recoded multiplier requires the use of the system's adder to add to AC either the positive outputs of the ICAND register or the complement outputs.* We may also observe that any digit pair $\cdots \bar{1} \ 1 \cdots$ is equivalent to $\cdots 0 \ \bar{1} \cdots$ and $\cdots 1 \ \bar{1} \cdots = \cdots 0 \ 1 \cdots$. Thus

$$\bar{1} \ 1 \ 0 \ \bar{1} \ 1 \ 0 \ 0 \ \bar{1} = 0 \ \bar{1} \ 0 \ 0 \ \bar{1} \ 0 \ 0 \ \bar{1}$$

and

$$1 \ \bar{1} \ 1 \ 0 \ \bar{1} \ 1 \ \bar{1} \ 0 = 1 \ 0 \ \bar{1} \ 0 \ \bar{1} \ 0 \ 1 \ 0$$

Each of these new versions (on the right) has fewer nonzero digits than its source and therefore requires fewer uses of the adder. Note that the $\bar{1} \ 1$ pairs occurred when an

---

*Here, and in subsequent pages, we assume that whenever we wish to subtract the ICAND, we add its 2's complement. The ICAND is assumed to be available in a flip-flop register, so that the 1's complement is available at the complement outputs of the flip-flops. When a subtraction of the ICAND is needed, this 1's complement may be gated to the adder, together with a 1 as a "carry" into the least significant digit to convert it to a 2's complement number.

isolated zero intruded in an otherwise continuous sequence of 1's and that replacing the $\bar{1}$ 1 pair produced a single $\bar{1}$ at the location of the zero. That is,

$$\underbrace{1\ 1\ 1\cdots1\ 0\ 1\ 1\cdots1}_{k\ 1\text{'s}} = 2^k - 2^j - 2^0$$

with a zero in position j

$$= \underbrace{1\ 0\ 0\ 0\cdots0\ \bar{1}\ 0\cdots0\ \bar{1}}_{}$$

$k + 1$ digits, all zero except
leftmost $= +1$, jth $= \bar{1}$ and
rightmost $= \bar{1}$

Similarly, a 1 $\bar{1}$ pair occurs whenever an isolated 1 appears in an otherwise continuous sequence of zeros, and to minimize adder uses we should not recode such a 1 but should allow it to stand.

Our rules now must be modified according to whether or not the scanning process has detected the start of a sequence of 1's in the multiplier. We start the scan with this "mode" indicator $= 0$, set the mode to 1 when a string of 1's starts, and return it to 0 when the string ends. A string ends only with two successive zeros. The rules are shown in Table 4.5.

TABLE 4.5   MODIFIED BOOTH RECODING[a]

	d_{j+1}	d_j	Action in position j
Mode indicator $= 0$			
These three actions are equivalent to doing no recoding of the IER if we have not detected the start of a sequence of 1's	0	0	None
	0	1	Add ICAND
	1	0	None
	1	1	Subtract ICAND and set mode indicator $= 1$
Mode indicator $= 1$			
These three actions are equivalent to replacing a sequence of 1's by an initial $\bar{1}$ followed by 0's, except that an isolated 0 in such a string is replaced by $\bar{1}$	0	0	Add ICAND and set mode indicator $= 0$
	0	1	None
	1	0	Subtract ICAND
	1	1	None

[a]Also called "canonical" recoding.

As always, following each digit-pair inspection the double-length register (AC, MQ) is shifted one position to the right. The modified procedure starts by inspecting $d_1 d_0$ (not $d_0 0$ as in the unmodified case) and concludes by inspecting $d_{n-1} d_{n-1}$; that is, at the last step the digit pair inspected is the sign digit and an assumed copy of it on the left.

It is not hard to show that on the average this modified procedure converts the multiplier to a form that has twice as many zeros as nonzero digits (see, for example, Ref. 3). Zero digits require only a shift operation and need not use the adder, and therefore this form of recoding improves the speed of multiplication as well as solving the problem of sign determination. Since random binary numbers have on the average as many zeros as ones, this recoding saves one-third of the uses of the adder, in a typical multiplication.

An important aspect of the recoding rule described in Table 4.5 is that it yields a minimal recoding (i.e., one having the smallest number of nonzero digits), and that among such minimal recodings it forms the unique one in which nonzero digits are separated by at least one zero. For this reason it is often referred to as *canonical* recoding. (We shall have further occasion to examine minimal recodings in our consideration of division procedures in Chapter 5.) As examples of minimal recodings, the following numbers are equal to one another and are minimally coded, but in only one are nonzero digits separated by at least one zero.

$$0 \ 1 \ 0 \ 1 \ 0 \ 1 \ 1 \ 0$$
$$0 \ 1 \ 0 \ 1 \ 1 \ 0 \ \bar{1} \ 0$$
$$0 \ 1 \ 1 \ 0 \ \bar{1} \ 0 \ \bar{1} \ 0$$
$$1 \ 0 \ \bar{1} \ 0 \ \bar{1} \ 0 \ \bar{1} \ 0$$

In performing the shift step in the modified Booth algorithm, we must be careful that the sign bit of the accumulator has the proper value after a shift. If the accumulator content represents a negative number, we both shift the 1 that is in the sign position and copy it again into the sign position. However, if a 1 appeared in the sign position merely as a result of adding positive quantities into the accumulator, then when this 1 is shifted, it must be replaced by a zero in the sign position. We may distinguish the following cases:

1. *IER and ICAND both positive*: Insert a 1 in the accumulator sign position when shifting across 1's, that is, when the mode = 1; otherwise, insert 0 in the sign position.
2. *IER positive and ICAND negative*: Copy the sign bit (whether 0 or 1) as well as shifting it, for either mode.
3. *IER negative and ICAND positive*: Follow the same procedure as for case 1.
4. *IER and ICAND both negative*: Follow the same procedure as for case 2.

In summary, whenever the ICAND is positive and not equal to zero we should insert a 1 in the sign position after each shift when shifting across 1's (i.e., when the mode = 1), but when shifting across 0's we insert a 0. If the ICAND is negative, the

TABLE 4.6

ICAND = 0101

IER = 0111

	Mode	AC	MQ		
Initially	0	0000	0111	1.1	$d_1d_0 = 11$ and mode = 0
	1	1011	0111		Subtract ICAND and set mode = 1
	1	1101	1011	1.2	Shift, and set sign = 1
				2.1	$d_2d_1 = 11$ and mode = 1 Neither add nor subtract
	1	1110	1101	2.2	Shift, and set sign = 1
				3.1	$d_3d_2 = 01$ and mode = 1 Neither add nor subtract
	1	1111	0110	3.2	Shift, and set sign = 1
		0101		4.1	$d_3d_3 = 00$ and mode = 1
	0	0100	0110		Add ICAND and set mode = 0
Sign of result ⟶		0010	0011	4.2	Shift

Result

sign bit is both shifted and copied back into the sign position for either mode. If either operand is zero, we would ordinarily detect this condition at the start and form a zero product at once, without executing the algorithm. (Note that if the ICAND is zero, executing this algorithm would give an erroneous result if the mode bit is set to 1 at any point.)

Some examples employing modified Booth recoding are given in Tables 4.6 and 4.7.

TABLE 4.7

ICAND = 1011

IER = 0111

	Mode	AC	MQ		
Initially:	0	0000	0111	1.1	$d_1d_0 = 11$ and mode = 0
	1	0101	0111		Subtract ICAND and set mode = 1
	1	0010	1011	1.2	Shift, copying sign
				2.1	$d_2d_1 = 11$ and mode = 1 Neither add nor subtract
	1	0001	0101	2.2	Shift, copying sign
				3.1	$d_3d_2 = 01$ and mode = 1 Neither add nor subtract
	1	0000	1010	3.2	Shift, copying sign
				4.1	$d_3d_3 = 00$ and mode = 1
	0	1011	1010		Add ICAND and set mode = 0
Sign of result ⟶		1101	1101	4.2	Shift

Result

4.4.4 Overlapping Triplets

In the preceding algorithm, the number of shifts performed between uses of the adder depends on the length of strings of 0's in the recoded multiplier. As MacSorley has pointed out [4], in some applications it may be preferable to use shifts of the same size at every step so that the total number of cycles is always the same. MacSorley describes an algorithm of that type, which examines the multiplier digits in groups of three, each such triplet overlapping its predecessor in one position. The algorithm is the modern descendant of the "shortcut" multiplication scheme that was used with mechanical desk calculators (now almost museum pieces) in which multiplication by each decimal digit x_j required x_j additions of the multiplicand. For $x_j > 5$, less time is required if $10 - x_j$ subtractions are performed in the x_j position and if x_{j+1} is then augmented by 1 before being multiplied. For example, the multiplier 293728 would be treated as 3 $\bar{1}$ 4 $\bar{3}$ 3 $\bar{2}$, requiring only 16 add-like operations instead of the 31 needed for the original form of the multiplier. To state the recoding rules somewhat more generally and more precisely, we define:

$$x_j = j\text{th digit of the multiplier in conventional form}$$

$$x_j^{(R)} = \text{recoded } j\text{th digit of the multiplier}$$

$$r = \text{an even-valued radix}$$

The recoding rules are:
For all j,

$$x_j^{(R)} = \begin{cases} x_j & \text{if } x_j < r/2 \\ -(r - x_j) & \text{if } x_j \geq r/2 \end{cases}$$

If any $x_j \geq r/2$, add 1 unit to $x_{j+1}^{(R)}$.

Note that when we add a unit to position $(j + 1)$, we add it to the recoded digit in that position rather than adding it to x_{j+1} and then recoding the sum. Either procedure is valid, but the foregoing rules have the feature that all digits may be simultaneously recoded, since there is no propagation of the unit addition.

Let us examine these rules for the useful case of $r = 4$. Our multiplication procedure uses x_j as it stands, if $x_j = 0$ or 1, and uses $-(4 - x_j)$ if $x_j = 2$ or 3, with the $(j + 1)$th position being augmented by 1 in these latter two cases. The actions are thus as shown in Table 4.8. For instance, from these recoding rules we may verify the following equivalences:

$$213 = 1\bar{2}2\bar{1}$$

$$1021 = 11\bar{2}1$$

$$2123 = 1\bar{2}2\bar{1}\bar{1}$$

$$1012 = 102\bar{2}$$

TABLE 4.8

x_j	Action in jth position	Compensating action in $(j+1)$th position
0	None	None
1	Add 1 $\times$ ICAND	None
2	Subtract 2 $\times$ ICAND	Add 1 $\times$ ICAND
3	Subtract 1 $\times$ ICAND	Add 1 $\times$ ICAND

Note that the recoded numbers employ the digit set

$$d_j = \{-2, -1, 0, 1, 2\}$$

This five-digit set constitutes a redundant representation for base 4 numbers, since it allows more than one way of representing a number with digits chosen from that set. Although we will consider redundant signed-digit number systems in more detail in Chapter 7, it may be noted here that the redundancy makes possible the independent recoding of the digit in each position by examining only that position and its immediately less significant neighbor. If, on the other hand, we had added the "carry" unit and then recoded the sum, we would have produced a nonredundant digit system which is subject to propagation effects. For example, 11113, recoded by this method, becomes $1\bar{2}\bar{2}\bar{2}\bar{2}\bar{1}$, whereas recoding *before* adding the carry unit would yield $1111\bar{2}\bar{1}$. We observe that the nonredundant digit set resulting from recoding *after* adding the carry is $d_j = \{-2, -1, 0, 1\}$.

Base 4 digits, of course, can be written as pairs of binary digits, and the procedure can easily be restated for pairs of bits in conventional binary numbers. Each digit pair in which a subtraction is performed requires addition of the ICAND into the position corresponding to the next more significant bit pair position. The table takes the form shown in Table 4.9. Since the compensating addition is required whenever the leftmost bit of the next lower-order pair is a 1, we can inspect 3 bits at a time, overlapping one with the next lower-order 3-bit group. Although we can inspect the multiplier from either end, it is more common to start from the right (i.e., the low-order end), examining the least significant digit pair $x_1 x_0$ and an assumed 0 to its right, then $x_3 x_2 x_1$, and so on, as summarized in Table 4.10.

TABLE 4.9

x_{j+1}	x_j	Action in jth bit position	Compensating action in $(j+2)$th bit position
0	0	None	None
0	1	Add 1 $\times$ ICAND	None
1	0	Subtract 2 $\times$ ICAND	Add 1 $\times$ ICAND
1	1	Subtract 1 $\times$ ICAND	Add 1 $\times$ ICAND

TABLE 4.10

Current digit pair		Left digit of previous pair,	
x_{j+1}	x_j	x_{j-1}	Action in position j
0	0	0	None
0	0	1	Add ICAND
0	1	0	Add ICAND
0	1	1	Add $2 \times$ ICAND
1	0	0	Subtract $2 \times$ ICAND
1	0	1	Subtract ICAND
1	1	0	Subtract ICAND
1	1	1	None

After each action, a right shift of (AC, MQ) by two positions is performed.* Thus the algorithm always requires $(n/2)$ shift steps instead of the n or $(n-1)$ shift steps needed in the conventional multiplication algorithm.

As a simple example, here is a 4-bit multiplication with ICAND = 0101 and IER = 1010. Note that we need an extra bit position to the left of AC, since we must be able to add $\pm 2 \times$ ICAND, and we also provide an extra position to the right of MQ to hold the leftmost bit of the previous pair inspected. This extra position to the right is initially set to 0.

ICAND = 0101

```
              AC              MQ
Extra bit                                  ┌─── Extra bit at right
at left      0 │0 0 0 0│    │1 0 1 0│ 0
               1 0 1 1 0                ╰──── Subtract 2 × ICAND

               1 0 1 1 0     1 0 1 0

               1 1 1 0 1     1 0│1 0 1    Shift right two positions
               1 1 0 1 1              ╰───── Subtract ICAND

               1 1 0 0 0     1 0 1 0 1

                 1 1 1 0     0 0 1 0 1    Shift right two positions
               └─────────────────────┘
                        Result
```

*We have assumed that the multiplier has an even number of bits, including its sign. If it has an odd number of bits, the leftmost triplet consists of the sign bit (and its copy) and the leftmost bit of the next right triplet. A one-position shift is performed after this triplet's action is performed.

Note that after the final shift the leftmost 2 bits of AC are sign bits. Since we performed $n/2$ shifts of two positions, the $2n_{bit}$ double-length register contains 2 sign bits at the left, followed by $2(n - 1)$ nonsign bits, as in Fig. 4.2b.

Note also that since the accumulator and the adder were extended one position to the left to accommodate $\pm 2 \times$ ICAND, we must extend the sign of $\pm 1 \times$ ICAND one position to the left whenever we add that quantity. An instance of this occurs in the foregoing example in the line labeled "Subtract ICAND."

Alternatively, one may take the point of view that the triplet being examined is the current pair of bits and the low-order bit of the next-higher-order pair. In Table 4.10 we would regard $x_j x_{j-1}$ as the current pair and x_{j+1} as the low-order bit of the next-higher-order pair. The actions to be taken are the same as those of Table 4.10, but since position $(j - 1)$ is now our reference, the quantities we add or subtract in that position must be twice the magnitude of those we add or subtract using position j as our reference. Table 4.11 shows the actions in this case.

Here is our previous example as it would be executed using Table 4.11. Note that we do not need the extra bit on the right to start our procedure, but when we examine the leftmost digit pair, we form the third member of that final triplet by copying the left bit of the pair. That is, with an even number of bits in the IER, either the rightmost pair or the leftmost pair must be extended by another bit, depending on whether we use the interpretation in Table 4.10 or that in Table 4.11.

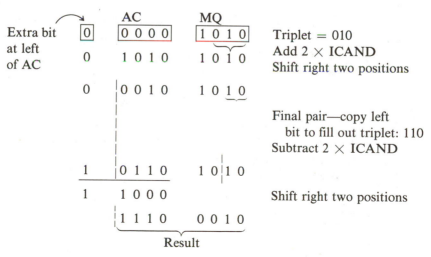

$$ICAND = 0101$$

$$2 \times ICAND = 01010$$

$$-2 \times ICAND = 10110$$

It is not difficult to modify this algorithm to inspect groups of 4 bits and to perform shifts of three positions. The reader is urged to confirm his understanding of the procedure by making this modification.

TABLE 4.11

x_{j+1}	x_j	x_{j-1}	Action in position $(j-1)$
0	0	0	None
0	0	1	$+2$
0	1	0	$+2$
0	1	1	$+4$
1	0	0	-4
1	0	1	-2
1	1	0	-2
1	1	1	None

$$\left. \begin{array}{c} +4 \\ -4 \\ -2 \\ -2 \end{array} \right\} \times \text{ICAND}$$

4.4.5 Overlapping Triplets with Magnitude Numbers

Although the overlapping triplets method has been presented in the preceding pages as a development of the Booth recoding scheme for multiplying numbers that may be in 2's-complement form, it can also be applied to numbers in magnitude form, since it is in essence simply a recoding of the multiplier into a signed-digit representation that is fully equivalent to the original multiplier. However, for n-bit numbers in magnitude form, this equivalence may require $n/2 + 1$ digits in the recoded number, while 2's-complement numbers can always be expressed in $n/2$ recoded digits. (We assume n an even number.) This is so because in 2's-complement numbers the leftmost bit d_{n-1} represents the sign bit, and if that bit is 1, the leftmost nonzero digit in the recoded form will be negative (-1 or -2) and the compensating $+1$ digit at the left end of the sequence is not needed. However, with numbers in the magnitude form, $d_{n-1} = 1$ represents a positive quantity, and since the recoding in that position gives a negative digit, the positive carry to the left is now necessary.

For example, consider the digit group

$$d_3 d_2 d_1 d_0 = 1010$$

As a 2's-complement number, this group's value is -6, but as a magnitude it represents $+10$, and our recoding should show these differences. In the complement case, we ignore the compensating addition of 1 in the position to the left of d_3, but in the magnitude case we include it, as shown below.

1. $\overbrace{1\ 0}\ \underbrace{1\ 0}\ |\ 0 = -6$

    $\quad\ \ \bar{1}\quad\ \ \bar{2}\quad\ = -1 \times 4^1 - 2 \times 4^0 = -6$

2. $\underbrace{0\ 0}\ |\overbrace{1\ 0}\ \underbrace{1\ 0}\ |\ 0 = +10$

    $\qquad\ \ 1\ \ \quad \bar{1}\quad\ \ \bar{2}\quad = 1 \times 4^2 - 1 \times 4^1 - 2 \times 4^0 = +10$

Here is another example:

$$d_7 d_6 d_5 d_4 d_3 d_2 d_1 d_0 = 1\ 1\ 1\ 1\ 0\ 1\ 1\ 0$$

which represents -10 as a 2's-complement number or $+246$ as a magnitude. The corresponding recodings are:

3. $\overbrace{1\ 1}\ \overbrace{1\ 1}\ 0\ \overbrace{1\ 1}\ 0 | 0$

      $0\quad \bar{1}\quad 2\quad \bar{2}\ \ = -1 \times 4^2 + 2 \times 4^1 - 2 \times 4^0 = -10$

4. $0\ 0 | \overbrace{1\ 1}\ \overbrace{1\ 1}\ 0\ \overbrace{1\ 1}\ 0 | 0$

      $1\quad 0\quad \bar{1}\quad 2\quad \bar{2}\ \ = 1 \times 4^4 - 1 \times 4^2 + 2 \times 4^1 - 2 \times 4^0 = 246$

To perform operations on magnitude numbers, we extend the accumulator two positions to the left, to accommodate $2 \times$ ICAND and also a sign digit. For instance, to multiply $X = 1010$ by $Y = 1010$, we proceed as follows:

$$\text{ICAND} = 0\ 0\ 1\ 0\ 1\ 0$$

$$-\text{ICAND} = 1\ 1\ 0\ 1\ 1\ 0$$

$$2 \times \text{ICAND} = 0\ 1\ 0\ 1\ 0\ 0$$

$$-2 \times \text{ICAND} = 1\ 0\ 1\ 1\ 0\ 0$$

Left extension	AC	MQ	Right extension
0 0	0 0 0 0	1 0 1 0	0
1 0	1 1 0 0	1 0 1 0	0 Subtract $2 \times$ ICAND
1 1	1 0 1 1	0 0 1 0	1 Shift right two positions
1 1	0 1 1 0		Subtract $1 \times$ ICAND
1 1	0 0 0 1	0 0 1 0	1
1 1	1 1 0 0	0 1 0 0	1 Shift right two positions
0 0	1 0 1 0		Add $1 \times$ ICAND
0 0	0 1 1 0	0 1 0 0	

Result

4.5 SPEEDING UP MULTIPLICATION

Unless we resort to table-lookup methods, which are very expensive, multiplication requires the addition of many operands. If only a single adder is available, it must be used many times, and there is a limited number of ways in which we can speed up the addition or reduce the necessary uses of the adder. With many adders available, however, many more methods for fast multiplication suggest themselves.

4.5.1 Deferred Carry Assimilation: A Single-Adder Method

In the adder structure of Fig. 4.1, each stage's carry digit is connected as an input to the next stage, and carries must propagate to completion before another addition can take place. However, if we provide another register to save the carry digits temporarily, we may add them in at the time of the next addend's arrival, postponing the final full-length assimilation just as we did in the adder structure of Fig. 3.13. The structure is shown in Fig. 4.3.

As usual, the digits of the IER are inspected and the ICAND is accordingly added or not. However, the resulting carry is now held in the carry-save register. Then (AC, MQ) is shifted right and another addition takes place. Because of the right shift of the partial product, the saved carry is properly aligned for addition at the next step. Each addition takes only the time required for an adder stage to act, since the adder stages are effectively in parallel, with no sequential propagation of carries. Only after the last conditional addition of the ICAND is it necessary to assimilate completely the

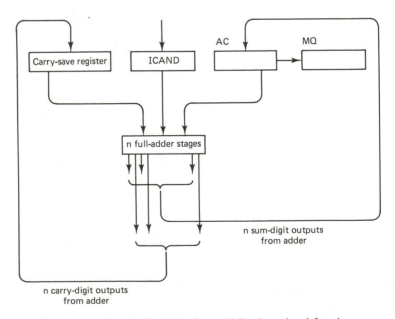

Figure 4.3   Register configuration for multiplication using deferred carry assimilation.

contents of the carry-save register, and at this moment carry look-ahead circuitry can be gated into action to perform the final assimilation rapidly.

Note that the deferred carry assimilation (abbreviated DCA) method cannot be further accelerated by the use of Booth recoding, since the n full-adder stages are used at every step, whether or not the multiplier digit is 0. We may use Booth recoding here merely as a way of handling the problem of a negative multiplier. However, even this is not necessary, since the DCA method is well suited to implementation of Robertson's method for handling signs. DCA gives the correct form for the result if the IER is positive, and for a negative IER we simply complement both operands at the start or else we subtract the ICAND at the last step just before assimilating the carry vector.

Of course, all of our add-and-shift algorithms for multiplication (the Booth ones as well as the conventional one) may proceed from the most significant end of the multiplier instead of from its low-order end. We then perform additions at the right end of the partial product and shift the partial product to the left. Now, however, we must provide circuitry to allow a possible carry to propagate beyond the n-digit addition at each step to the leftmost bit of the partial product. The n-bit accumulator needs an n-bit left extension, which cannot be used for the multiplier digits because of this possibility of a carry. For instance, if ICAND = 1111 (magnitude) and IER = 1001 (magnitude), scanning the IER from the left would yield

MD		
$\boxed{1111}$		

MQ	Double-length AC	
$\boxed{1001}$	$\boxed{00000000}$	Initially
	00001111	$MQ_3 = 1$, so add ICAND
0010	00011110	Shift left
0100	00111100	$MQ_3 = 0$, so shift left
1000	01111000	$MQ_3 = 0$, so shift left
	$+\ 1111$	$MQ_3 = 1$, so add ICAND
	10000111	

Result

The final addition step produces (in this case) a carry that propagates four bit positions to the left of the positions in which the ICAND addition occurs, and it is the possibility of such a carry that accounts for the additional circuit complexity of this method. It is interesting to observe that it is not possible to have a carry that propagates over the entire length of the $2n$-bit accumulator. The longest carry chain that can occur is still only n bit positions, so scanning from the left incurs no time penalty for carry propagation.

4.5.2 Arrays for Multiplying

An array of half-adders and full-adders can be constructed to add simultaneously all the terms $x_j \times$ ICAND. Consider a 4-bit $\times$ 4-bit multiplication, in which we must add four numbers as below:

$$A = x_0 \times \text{ICAND}$$
$$B = x_1 \times \text{ICAND}$$
$$C = x_2 \times \text{ICAND}$$
$$D = x_3 \times \text{ICAND}$$

The array of Fig. 4.4 will perform this multiplication. The row of half-adders at the top accepts $x_0 \times$ ICAND and $x_1 \times$ ICAND, producing a carry vector and a sum vector. The carry vector is connected one position to the left in the second row, where it is added, in a row of full adders, to the first row's sum vector and to $x_2 \times$ ICAND. Another $x_j \times$ ICAND is added in each succeeding row, and the bottom row assimilates the carry. Since the product digits p_j are not all formed at the same time, the

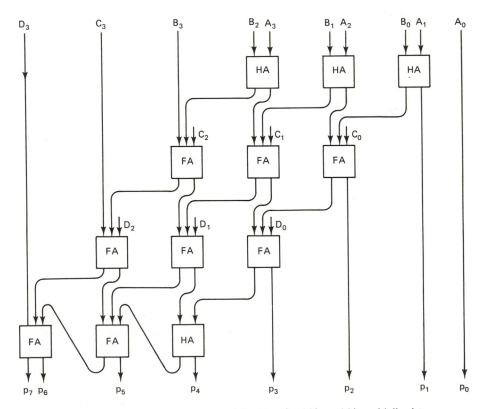

Figure 4.4 Array of half-adders and full-adders for 4-bit $\times$ 4-bit multiplication.

input levels must be maintained long enough to allow all outputs to reach their final values.

An n-digit multiplier of this type requires n half-adders and $(n - 1)^2 - 1$ full-adders, as well as the gating circuitry needed to admit all the $x_j \times$ ICAND terms to their proper places. The longest propagation path is through the rightmost adders in each row and then across the carry assimilation row at the bottom. This path contains $2(n - 2)$ full-adders and two half-adders, making the delay time for a complete multiplication be about $2n - 3$ full-adder times.

The array of Fig. 4.4 can be modified for pipelining by interposing latch registers between the rows of adders, just as we did in the addition array of Chapter 3. Pipelining allows overlapping of a succession of multiplications. [Note that the carry-assimilation row would have to be replaced by $(n - 1)$ successive rows in a pipelined version.]

Although the structure of Fig. 4.4 functions correctly for multiplication of two magnitudes, it does not give correct results for negative numbers in 2's-complement form. The difficulty is similar to those that we saw in the first algorithm of this chapter, where we found that each addition of a negative multiplicand required that its sign digits be extended left to the sign position of the final product.

Just as in the sequential add-and-shift algorithm there are several ways of handling negative numbers in a multiplication array. One obvious way is simply to treat all IER and ICAND digits as positive, using an array such as that of Fig. 4.4, and to add in any required correction terms $2(1 - |X|)$ or $2(|X| + |Y|)$ in extra rows of the matrix. This procedure is not satisfactory because it requires more array components and more time than needed, and we can devise better ways. We examine first a technique due to Baugh and Wooley [5]. As before we take ICAND $Y = y_{n-1} y_{n-2} \cdots y_1 y_0$ and IER $X = x_{n-1} x_{n-2} \cdots x_1 x_0$, where each operand is either positive, or in 2's-complement form if negative, and y_{n-1} and x_{n-1} are the sign digits. The true values of X and Y are therefore:

$$Y = -y_{n-1} \times 2^{n-1} + \sum_0^{n-2} y_i \times 2^i$$

$$X = -x_{n-1} \times 2^{n-1} + \sum_0^{n-2} x_i \times 2^i$$

where we have taken both numbers to be integers. The partial products to be added in the rows of the multiplication array are thus as shown in Fig. 4.5.

The two circled digit groups are each either 0, if y_{n-1} or x_{n-1}, respectively, is 0, or are negative, if the corresponding sign digit y_{n-1} or x_{n-1} is 1. That is, if $y_{n-1} = 1$, we must subtract the nonsign digits of X, in the positions shown, and correspondingly, if $x_{n-1} = 1$, we must subtract the nonsign digits of Y, in the positions shown. To maintain uniform structure of our array, so that we use adder units throughout, we can instead add the 2's complement of each of these quantities, which we form in the usual way by adding a unit to the 1's complement.

The digit pattern thus appears as in Fig. 4.6. Note that the digit pair $x_{n-1} x_{n-1}$ at the left forms 1 1 if X is negative and 0 0 otherwise; that is, they form a left exten-

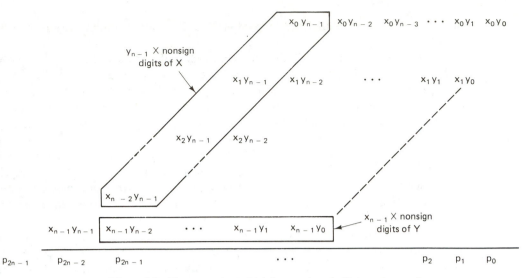

Figure 4.5   Bit pattern in multiplying numbers in 2's-complement form.

sion of the sign bit; $y_{n-1}y_{n-1}$ functions similarly for Y. We can form a somewhat simpler pattern at the left by noting the equivalence of the patterns

$$
\begin{array}{ccccc}
x_{n-1} & x_{n-1} & & 0 & \overline{x}_{n-1} \\
y_{n-1} & y_{n-1} & \text{and} & 0 & \overline{y}_{n-1} \\
\hline
& x_{n-1} \cdot y_{n-1} & & 1 & x_{n-1} \cdot y_{n-1}
\end{array}
$$

when subjected to columnwise addition. The digit pattern to be added then takes the form shown in Fig. 4.7.

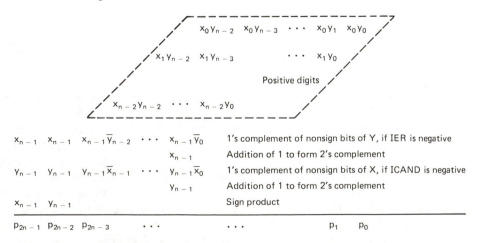

Figure 4.6   Rearrangement of the bit pattern of Fig. 4.5.

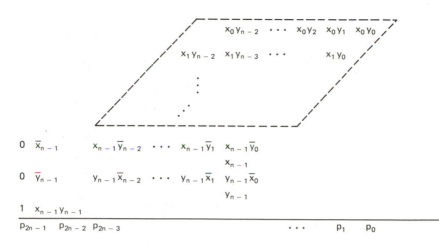

Figure 4.7   Final rearrangement of Fig. 4.5, using the method of Baugh and Wooley.

Figure 4.8 shows an array for multiplication of 4-bit numbers by this method. Since the array accepts two operands each having a sign and $(n-1)$ nonsign bits, the result has two sign digits, p_{2n-1} and p_{2n-2}, and $2 \times (n-1)$ nonsign digits. The sign digits will be 0 0 for positive products or 1 1 for negative products, except for the special case of $X = Y = 10 \cdots 0$, in which case the result will be $0100 \cdots 0$. That is, this structure treats operands $10 \cdots 0$ as being negative numbers of maximum magnitude, so that in the 4-bit structure, we have $1000 \times 1000 = 01000000$, that is $-8 \times -8 = +64$, just as we observed in Fig. 4.2b. We may observe also that the structure requires the operand digits to be available in both complemented and uncomplemented forms, but this requirement is easily met in typical electronic circuits, especially if the bits are held in flip-flops, which have both signals available.

An alternative method for handling 2's-complement operands is to provide a more versatile array cell, one which can add or subtract under external command, and then to use simple Booth recoding of the multiplier, so that each succeeding digit group is either added or subtracted or not admitted to the row of cells. Several such circuits have been described in the literature [6,7].

Yet another method is exemplified in the array structure of Pezaris [8], which uses elements that perform addition (in the central trapezoidal portion of the array) and elements that accept a subtractive input in the bordering portions of that array.

4.5.3 Wallace and Dadda Multipliers

Besides the array multiplier of the preceding, section, there are several other methods of adding the various $x_j \times$ ICAND rows to form the product. We consider first the method due to C. S. Wallace [9], in which groups of three of the $x_j \times$ ICAND vectors are applied as inputs to banks of (3, 2) adders, each such triplet of vectors producing two vectors at the outputs. Triplets of these output vectors are then applied as inputs

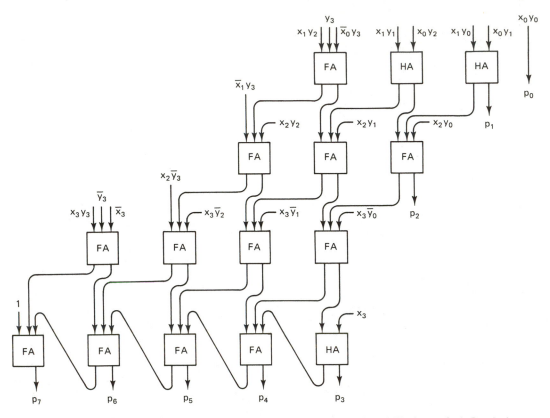

Figure 4.8 4-bit $\times$ 4-bit matrix for multiplication by Baugh and Wooley method. Operands either positive or in 2's-complement form.

to the next level of adders, and so forth. Figure 4.9 shows the structure of an 8-bit $\times$ 8-bit circuit.

Units 1 and 2 in this structure are identical, each receiving a set of three 8-bit vectors. As shown below, these vectors are added in a set of six (3, 2) adders (i.e., full-adders) and two (2, 2) adders (half-adders). The rightmost and leftmost digits pass straight through.

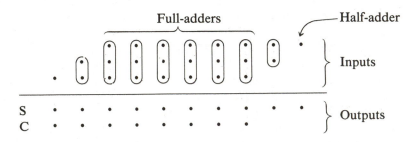

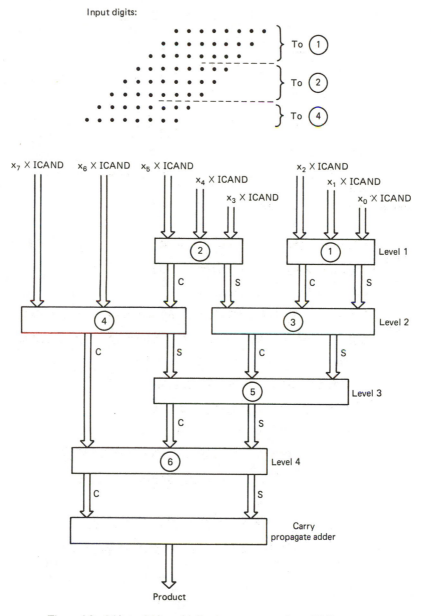

Figure 4.9   8-bit $\times$ 8-bit multiplication structure using a Wallace tree.

Unit 3 receives the S and C outputs of unit 1 and the S output of unit 2:

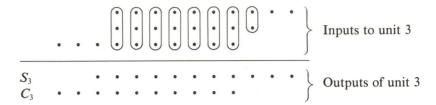

Unit 4 has as inputs the C output of unit 2 and the vectors $x_6 \times$ ICAND and $x_7 \times$ ICAND which were left over in forming the vector triplets at level 1:

We may continue in this fashion to generate inputs and outputs of the various units. The number of adders used is tabulated below.

Unit number	Number of full-adders	Number of half-adders
1	6	2
2	6	2
3	7	1
4	6	2
5	6	4
6	7	4
Total	38	15

We may note several interesting points about this Wallace tree multiplier. First, it is a form of deferred carry assimilation structure, since not until the last level, where only two vectors remain, are the carries absorbed. Second, since several low-order digits are in their final correct form at that point, the worst-case carry propagation is over substantially less than the double-length digit group. Third, the number of levels required is given approximately by

$$k = \frac{\log_2 n - 1}{\log_2 3 - 1}$$

(This follows from the fact that V_O, the number of output vectors from a given level, is related to V_I, the number of input vectors to that level, by

$$V_O = \left\lceil \frac{2}{3} \times V_I \right\rceil$$

which leads to

$$\underbrace{\left\lceil \left\lceil \left\lceil \frac{2}{3} \times n \right\rceil \times \frac{2}{3} \right\rceil \times \frac{2}{3} \right\rceil \times \cdots \frac{2}{3} \right\rceil}_{k \text{ factors}} = 2)$$

Fourth, the number of full-adders used is the order of n^2. This, of course, is not a tight bound on the number of adders—a closed-form expression for the number of adders is too cumbersome to be useful. However, a careful count of the number of adders needed for several values of n is tabulated below.

n	Number of full-adders	Number of half-adders	Total	Number of levels
6	16	12	28	3
8	38	15	53	4
12	102	29	131	5
16	192	56	248	6
24	488	87	575	7

An improvement on the Wallace structure has been made by L. Dadda [10], who has devised an arrangement that uses fewer adders, although the same number of levels are required. Dadda points out that since the last level produces two vectors (the inputs to the carry propagating adder), the preceding level had at most 3, its predecessor at most 4, and so on. This series—2, 3, 4, 6, 9, 13, 19, 28—is formed by the relation

$$T_{j+1} = \left\lfloor T_j \times \frac{3}{2} \right\rfloor$$

where T_j is the jth term in the sequence. To minimize the number of adder elements in the total structure, we start by using in the first level only enough full-adders and half-adders to reduce the initial n-row matrix to one having a number of rows equal to one of the terms in this series. Then at each succeeding level we use whatever number of adders we need to form the matrices having numbers of rows equal to the successively smaller numbers in this series. For example, a 12×12 multiplication could be represented by the dot pattern of Fig. 4.10, and with the full-adders and half-adders shown, it could be reduced at level 1 to a matrix with nine rows maximum. 9 is a member of our series, and level 2 would then use only enough adders to reduce the array to one

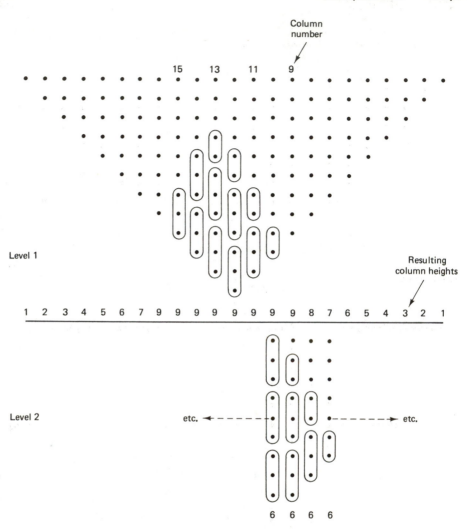

Figure 4.10 Using the Dadda method for 12-bit × 12-bit multiplication.

having six rows (since 6 is the next term in the series). Note that in Dadda's method, the adders are clustered initially toward the center, and only at the later stages are the low-order and high-order digits involved. Thus this method uses somewhat fewer adder units of each type than does the Wallace method. For example, in an 8-bit × 8-bit multiplication, the Dadda method will be found to require only 35 full-adders and 7 half-adders, while the Wallace arrangement requires 38 and 15, respectively. However, the Wallace structure produces final values of the low-order digits in early stages of the process, thus allowing the final assimilation of carries to be executed over somewhat shorter vectors than in the Dadda method.

How do we handle negative numbers in these structures? If negative numbers are in complement form, we can of course convert to sign and magnitude, carry out our multiplication, and recomplement if the result is negative. However, we prefer to handle our numbers directly in complement form, and this can be done at the expense of a slight increase in circuitry. Two methods are available:

1. Add in the required correction term, which is of the form $2 - 2X$, or
2. Execute a simple Booth recoding of the multiplier.

Recall (Section 4.4) that if the ICAND Y is negative and the IER X is positive, multiplication yields

$$(2 - |Y|) \times X = 2X - |Y| \times X$$

so to form the correct result $2 - |Y| \cdot X$ we must add a correction $2 - 2X$, which is $2(1 - X)$, that is, the 2's complement of the nonsign bits of X shifted left one position. This group (in 1's-complement form) is available as the $\overline{Q}$ outputs (i.e., the complement outputs) of the flip-flops of the register holding X and thus can easily be admitted as another vector of $(n - 1)$ bits at the left of either the Wallace or the Dadda structure. We also admit a bit at the right of the group to convert the 1's complement to 2's complement. The dot pattern for an 8-bit $\times$ 8-bit multiplication is then

8-bit correction term (1's complement and C_0 digit)

Clearly, somewhat more circuitry will be needed to sum this slightly larger dot pattern. If the multiplier X is the negative quantity and Y is positive, the correction term is $2(1 - Y)$, and if both factors are negative, we complement them both before multiplying, to avoid the need for correction.

Alternatively, we may use the simple Booth scheme for recoding the multiplier. In this method, d'_j, the new multiplier digit in position j, is a function only of d_j and d_{j-1}, and thus all digits of the recoded multiplier can be formed in parallel. (The modified Booth method, which reduces the number of nonzero digits in the multiplier, necessarily involves a sequential scan to detect strings of 1's and thus is not adaptable to a circuit in which we use all multiplier digits at the same time.) Each -1 in the recoded multiplier requires subtraction of the ICAND in the corresponding position, an operation that can be done by (1) admitting the $\overline{Q}$ outputs of the ICAND register instead of the Q outputs, (2) admitting copies of the sign bit in positions all the way to the left, and (3) entering a $C_0 = 1$ digit in the least significant position of the vector. A 4×4 dot pattern would be as follows:

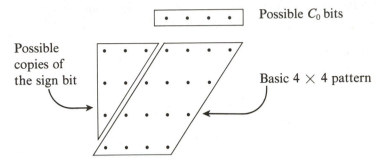

The basic $n \times n$ pattern must be augmented by $(n/2)(n-1)$ possible sign bits and n possible C_0 bits if we use this method. Thus the method entails substantially greater costs than the method of adding in a correction term.

It is of interest to note that more complex counters than the (3, 2) and (2, 2) adders in the foregoing examples will reduce the number of levels and therefore speed up the process. (7, 3) counters or (15, 4) counters could be very useful in accelerating multiplication of longer numbers, for example, 32-bit $\times$ 32-bit groups. Of course, it is possible to construct higher-order counters from simpler ones, if we are willing to accept the cost of increased time delay. For instance, a (7, 3) counter could be constructed from (3, 2) counters as in Fig. 4.11. The general procedure for constructing (N, M) counters from (p, q) counters starts with a dot pattern of N dots in one column and groups these into sets of p inputs, with possibly $N - \lfloor N/p \rfloor$ left over. Outputs of these counters are arranged in columns and the process is repeated. The dot patterns yielding Fig. 4.11 are as follows:

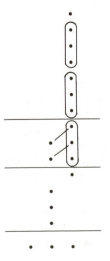

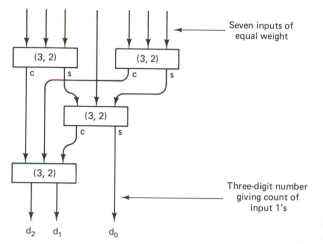

Figure 4.11   A (7, 3) counter built from (3, 2) counters.

The generalization follows easily, along the lines shown by the examples in Tables 3.8 and 3.9.

With higher-order counters available, an alternative approach to multiplication is to perform the operation in serial fashion [11], counting the number of bits in successive columns of the dot pattern, starting from the rightmost. As each such count is formed, its binary representation is added to the previous sum of counts, which must in the meantime have been shifted one position to the right. The elements of a structure for performing this operation are shown in Fig. 4.12. This circuit performs an $n \times n$ multiplication using an (n, q) counter, where $q = \lceil \log_2 n \rceil$. The product register had $2n + q - (1 \text{ or } 2)$ positions. After each column count is formed, it is added to the leftmost q positions of the product register. The sum is reinserted in those positions, and the product register is then shifted one position to the right. After all $(2n - 1)$ bit columns have been counted in, the product is in the rightmost $2n$ positions of the product register.

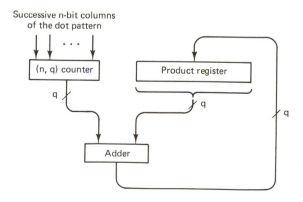

Figure 4.12   Structure for serial multiplication, accepting successive columns of the dot pattern.

As an example, consider ICAND and IER both equal to $1111 = 15_{10}$. The product register contents are as follows:

0 0 0|0 0 0 0 0 0	Initially
0 0 1|0 0 0 0 0 0	1. Add 1 = count of 1's in column 0
0 0 0|1 0 0 0 0 0	Shift right
0 1 0|1 0 0 0 0 0	2. Add 2 = count of 1's in column 1
0 0 1|0 1 0 0 0 0	Shift right
1 0 0|0 1 0 0 0 0	3. Add 3 = count of 1's in column 2
0 1 0|0 0 1 0 0 0	Shift right
1 1 0|0 0 1 0 0 0	4. Add 4 = count of 1's in column 3
0 1 1|0 0 0 1 0 0	Shift right
1 1 0|0 0 0 1 0 0	5. Add 3 = count of 1's in column 4
0 1 1|0 0 0 0 1 0	Shift right
1 0 1|0 0 0 0 1 0	6. Add 2 = count of 1's in column 5
0 1 0|1 0 0 0 0 1	Shift right
0 1 1|1 0 0 0 0 1	7. Add 1 = count of 1's in column 6
	Do not shift

$2n$-bit product

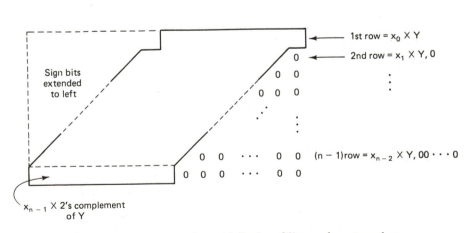

Figure 4.13 Dot pattern for multiplication of 2's-complement numbers.

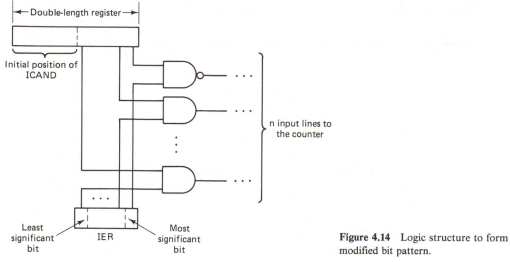

Figure 4.14   Logic structure to form modified bit pattern.

Since each iteration requires the sequential use of both the counter and the adder, the time delay of these components determines the rate at which the columns can be processed. Note also that there are $(2n - 1)$ columns to be processed.

In the simple form shown in Fig. 4.12, the structure can accept only magnitude numbers, and the proper sign for the result must be separately determined. A small modification, however, enables the structure to accept numbers in either positive or 2's-complement form. The digit pattern for multiplication of 2's-complement numbers can be displayed as shown in Fig. 4.13. We may replace the entire bottom row, which is

$$x_{n-1} \times \overline{Y} \; 0 \; 0 \; - - - \; 0 \; 0$$

by two rows that have the same sum as this one and that convert the 1's complement of Y to 2's-complement form:

$$\overline{x_{n-1}y_{n-1}} \; \overline{x_{n-1}y_{n-2}} \; - - - \; \overline{x_{n-1}y_0} \; \; 1 \; 1 \; - - - \; 1 \; 1$$
$$0 \qquad\quad 0 \qquad - - - \quad 0 \qquad 0 \; 0 \; - - - \; 0 \; 1$$

This modified bit pattern can be admitted to the counter one column at a time by the gating structure of Fig. 4.14. As the ICAND is shifted to the right in its double-length register, its sign bit is copied in at the left. The NAND gate forms the sequence of 1's at the right end of the first one of the two rows that we substituted. The second of those rows can be formed simply by initializing to 1 the first carry into the adder. Following the sequence of 1's the NAND gate forms the $\overline{x_{n-1}y_j}$ terms.

4.5.4 Two Examples of Large Fast Multipliers

An interesting combination of a multiplier recoding scheme and a Wallace adder tree was used in the IBM 7030 computer, known as STRETCH [12]. This machine used either 48-bit or 96-bit numbers, and it was thus convenient to retire 12 bits of the multiplier at each step, in the tree structure of Fig. 4.15.

      The groups of (3, 2) adders are arranged in the (by now) familiar Wallace structure. After the first 12 bits of the IER have been multiplied by the ICAND in adder groups 1, 2, 3, and 4, the resulting partial sum and partial carry groups are admitted to adder group 2 at the same time as the next more significant 12 bits of the IER, multiplied by the ICAND. Thus each product of 12 bits of the IER, multiplied by the ICAND, is admitted to the right end of the next-higher-order group. When all the IER digits have been processed, the partial sum and partial carry groups are gated into the carry assimilation adder.

      The recoding used in STRETCH was a 4-bit version of the recoding pattern of Table 4.11, as shown in Table 4.12. ± 2, ± 4, and ± 8 times the ICAND are easily

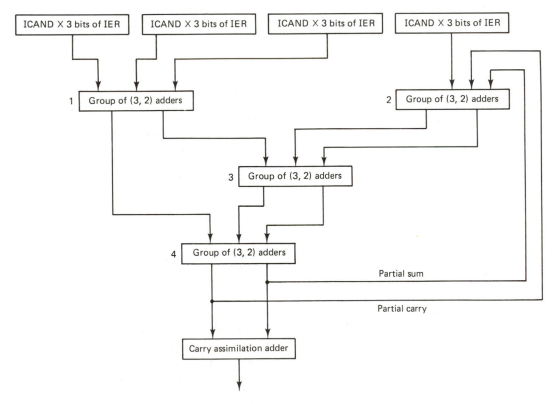

Figure 4.15   Wallace tree structure in STRETCH.

TABLE 4.12

Overlap bit	Current triplet			Action in
x_{j+3}	x_{j+2}	x_{j+1}	x_j	position j
0	0	0	0	None
0	0	0	1	+2
0	0	1	0	+2
0	0	1	1	+4
0	1	0	0	+4
0	1	0	1	+6
0	1	1	0	+6
0	1	1	1	+8
1	0	0	0	−8
1	0	0	1	−6
1	0	1	0	−6
1	0	1	1	−4
1	1	0	0	−4
1	1	0	1	−2
1	1	1	0	−2
1	1	1	1	None

(The actions +2 through −2 are grouped with $\times$ ICAND)

obtained by shifts of the ICAND or its complement. $\pm 6 \times$ ICAND requires a preliminary step and a temporary register. With a 48-bit multiplier, denoted $x_{47}x_{46} \cdots x_2 x_1 x_0$, the 4-tuples examined on the first pass are

$$x_3 \ x_2 \ x_1 \ x_0$$
$$x_6 \ x_5 \ x_4 \ x_3$$
$$x_9 \ x_8 \ x_7 \ x_6$$
$$x_{12} \ x_{11} \ x_{10} \ x_9$$

On the final pass, x_{47} is replicated on the left, and the 4-tuples are

$$x_{39} \ x_{38} \ x_{37} \ x_{36}$$
$$x_{42} \ x_{41} \ x_{40} \ x_{39}$$
$$x_{45} \ x_{44} \ x_{43} \ x_{42}$$
$$x_{47} \ x_{47} \ x_{46} \ x_{45}$$

A similar procedure was used in the multiplier circuits of the IBM 360 Model 91 [13]. Here also 12 IER bits were "retired" on each pass through the structure, but overlapping triplets were used (instead of 4-tuples), as summarized in Table 4.11. Therefore, six partial product terms were formed on each pass, and the Wallace tree was somewhat deeper, as shown in Fig. 4.16.

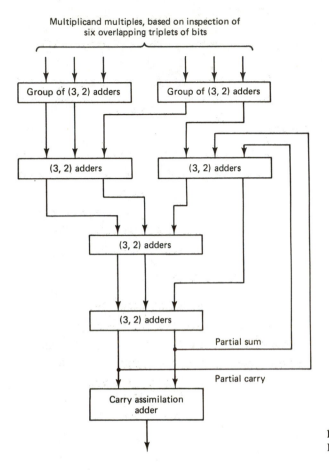

Multiplicand multiples, based on inspection of
six overlapping triplets of bits

Figure 4.16 Wallace tree structure in
IBM 360 Model 91.

REFERENCES

1. Robertson, J. E., Two's complement multiplication in binary parallel computers, *IRE Trans. El. Comp.*, vol. EC-4, no. 3, Sept. 1955, pp. 118–119.

2. Booth, A. D., A signed binary multiplication technique, *Quart. Jl. Mech. Appl. Math.*, vol. 4, pt. 2, June 1951, pp. 236–240.

3. Scott, N. R., *Electronic Computer Technology*, McGraw-Hill, New York, 1970, pp. 365–367.

4. MacSorley, O. L., High-speed arithmetic in binary computers, *Proc. IRE*, vol. 49, no. 1, Jan. 1961, pp. 67–91.

5. Baugh, C. R., and B. A. Wooley, A two's complement parallel array multiplication algorithm, *IEEE Trans. Comp.*, vol. C-22, no. 12, Dec. 1973, pp. 1045–1047.

6. Majithia, J. C., and R. Kitai, An iterative array for multiplication of signed binary numbers, *IEEE Trans. Comp.*, vol. C-20, no. 2, Feb. 1971, pp. 214–216.

7. Bandyopadhyay, S., S. Basu, and A. K. Choudhory, An iterative array for multiplication of signed binary numbers, *IEEE Trans. Comp.*, vol. C-21, no. 8, Aug. 1972, pp. 921–922.

8. Pezaris, S. D., A 40-ns 17-bit by 17-bit multiplier, *IEEE Trans. Comp.*, vol. C-20, no. 4, Apr. 1971, pp. 442–447.

9. Wallace, C. S., A suggestion for a fast multiplier, *IEEE Trans. El. Comp.*, vol. EC-13, no. 1, Feb. 1964, pp. 14–17.

10. Dadda, L., Some schemes for parallel multipliers, *Alta Frequenza*, vol. 34, no. 5, May 1965, pp. 349–356.

11. Swartzlander, E. A., Jr., The quasi-serial multiplier, *IEEE Trans. Comp.*, vol. C-22, no. 4, Apr. 1973, pp. 317–321.

12. Buchholz, W., ed., *Planning a Computer System*, McGraw-Hill, New York, 1962, pp. 210–211.

13. Anderson, S. F., J. G. Earle, R. E. Goldschmidt, and D. M. Powers, The IBM System/360 Model 91: floating-point execution unit, *IBM Jl. R & D*, vol. 11, no. 1, Jan. 1967, pp. 34–53.

EXERCISES

4.1. A common procedure for rounding off in multiplication of two numbers in sign-magnitude form is to add 1 to the least significant position of the high-order part if the most significant digit of the low-order part is 1. This procedure closely approximates the rounding mode known as "round to nearest," since the resulting high-order part will be the nearest neighbor of the unrounded number except in the special situation where the low-order part of the unrounded number is $10 \cdots 0$. Show that we can perform this rounding easily if we start with a single 1 bit in the accumulator instead of clearing the accumulator to all zeros. Where should this 1 bit be placed in AC?

4.2. Recode each of the multiplier groups below by each of the following methods.
1. Booth recoding
2. Modified Booth recoding
3. Overlapping triplets (Table 4.10)
4. Overlapping triplets (Table 4.11)

Then verify that the recoded form is equivalent to the given form.

Multipliers (either positive or 2's complement):
(a) 1.01101011   (b) 0.01010101
(c) 1.1110101   (d) 1.01010110

4.3. In the overlapping-triplets method of multiplication, we saw that it is necessary to extend the accumulator one bit to the left to accommodate $\pm 2 \times$ ICAND. Show that if we permit the group $10 \cdots 0$ to represent a negative number of maximum magnitude, we must extend the accumulator 2 bits to the left so as to form the sign bits 01 when this number is multiplied by itself.

4.4. Make a careful count of the number of (3, 2) adders and the number of (2, 2) adders required in a Wallace tree for multiplication of 12-bit numbers, to confirm (if possible) the figures of 102 and 29 given in the text. How many final correct low-order digits are formed at each level of the tree? What is the worst-case carry propagation delay in combining the final two vectors from the tree; that is, through how many (3, 2) adder stages might a carry propagate?

4.5. Shown is a block diagram of an 8-bit × 8-bit "tree-add" multiplier. The eight registers in the top row contain copies of the multiplicand. These are either admitted or not admitted to the ADD units (with proper relative shifts) according to the values of the eight multiplier digits. The adder outputs are stored in temporary registers $T_1 \cdots T_4$, from which they are added into $T_5 T_6$ and then into T_7, each time with proper relative shift. In general, for $2^k \times 2^k$ multiplication by this method:

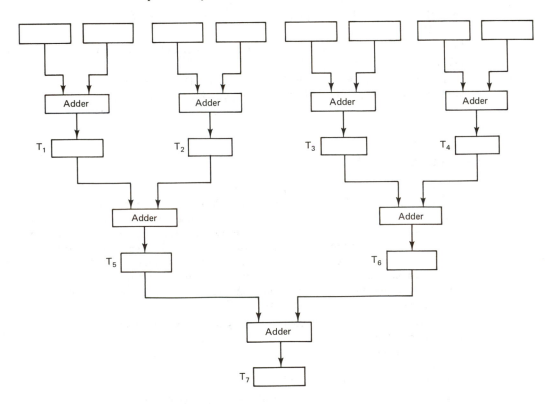

(a) How many add times are needed?

(b) How many temporary registers are needed, and of what length?

(c) How many (3, 2) adders and how many (2, 2) adders are needed to implement one of the ADD units in the top row? How many for the last ADD unit?

(d) Each of the addition operations in this structure requires the assimilation of carries. Compare the speed of this structure and the number of (3, 2) and (2, 2) adders that it uses with the corresponding figures for the Wallace tree of Fig. 4.9. Be sure to include the final carry assimilation step in the Wallace tree.

4.6. When we build (N, M) counters out of (p, q) counters, as shown in Fig. 4.11, we must use some care in interconnecting the (p, q) adders to ensure that digits of the same weight are being combined in each (p, q) adder. Which of the structures shown are satisfactory $(7, 3)$ counters?

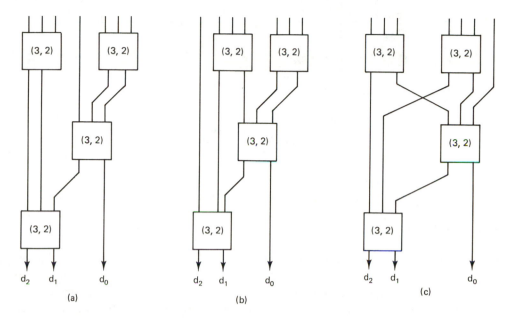

(a)   (b)   (c)

4.7. Suppose that it becomes practical to construct $(2^k - 1, k)$ counters for $k \leq 5$, with each counter having two gate delay times. If we use the optimum number and type of counters at each level, what would be the minimum number of levels needed in a Wallace tree multiplier of 64-bit $\times$ 64-bit numbers? (The last level will, of course, form two vectors.)

4.8. Trace through the signals formed in the matrix multiplier of Fig. 4.8 when $X = 1010$ and $Y = 1101$, to verify that the correct outputs are obtained. Modify that structure so that it can be used for pipelined multiplication, that is, for the overlapping multiplication of pairs $X_1 \times Y_1$, $X_2 \times Y_2$, and so on. What will be the total transit delay for the modified array; that is, how many adder delays are in the path? (Assume that full-adders and half-adders have the same delay.)

Division

5.1 INTRODUCTION

Division is the mathematical inverse of multiplication, and we may reasonably expect to find division algorithms that correspond closely to the step-by-step inverse of algorithms for multiplication. As we have seen, the usual multiplication algorithm accumulates, one after the other, the partial product terms formed by successive multiplications of the individual multiplier digits times the multiplicand, with intervening right shifts of the accumulated sum. In the corresponding division algorithm, we successively discover the digits of the quotient by trial subtractions of the divisor from the dividend, with intervening left shifts of the resulting remainder. We should expect that neither process will involve more than n steps (for n-digit factors), and we may also anticipate that just as there are methods for speeding up multiplication, there are also ways for speeding up division.

5.2 RESTORING AND NONRESTORING DIVISION

The central idea of our algorithm is easily stated: To determine how many times a divisor divides a dividend, we subtract the divisor from the dividend repeatedly until we encounter a negative result. The number of subtractions before overdrawing is the first digit of the quotient. In the common pencil-and-paper procedure, the subtractions are performed as far left as possible with respect to the numerator (the dividend). For example, to divide 2.2/9 we perform

$$
\begin{array}{r}
.244 \\
9\overline{\smash{)}2.200\ldots} \\
-1\ 8 \\
\hline
40 \\
-36 \\
\hline
40 \\
\vdots
\end{array}
$$

To do the same process by repeated subtraction, we perform

$$
\begin{array}{r}
22 \\
-\ 9 \\
\hline
13 \\
-\ 9 \\
\hline
4
\end{array}
\left.\vphantom{\begin{array}{r}22\\-9\\13\\-9\\4\end{array}}\right\}
\text{Two subtractions of the divisor before overdrawing}
$$

$$
\begin{array}{r}
-\ 9 \\
\hline
-\ 5
\end{array}
\left.\vphantom{\begin{array}{r}-9\\-5\end{array}}\right\}
\text{Third subtraction gives negative result (overdraft)}
$$

The first quotient digit is thus seen to be 2, although this fact was not evident until three subtractions had been performed.

We next "restore" our divisor to the negative remainder by adding it back in to get $+4$, and we then shift this positive remainder one position left to get 40, from which we again start our subtraction process. Proceeding in this manner, we "uncover" the successive quotient digits .2444 . . . .

In the restoring method of division, whenever we have overdrawn, we cancel out the last subtraction by adding the divisor in again before proceeding to the next digit. The nonrestoring method avoids this restoration of the divisor by shifting the negative remainder left and performing successive *additions* until the result becomes positive. We count the number of subtractions as a positive digit of the "pseudo-quotient" and the number of additions as a negative digit of the pseudo-quotient.

Thus

$$
\begin{array}{cccc}
\begin{array}{r}
22 \\
-\ 9 \\
\hline
-13 \\
-\ 9 \\
\hline
4 \\
-\ 9 \\
\hline
-\ 5
\end{array}
&
\begin{array}{r}
-50 \\
+\ 9 \\
\hline
41 \\
\cdot \\
\cdot \\
\cdot \\
+\ 4
\end{array}
&
\begin{array}{r}
+40 \\
-\ 9 \\
\hline
31 \\
\cdot \\
\cdot \\
\cdot \\
-\ 5
\end{array}
&
\begin{array}{r}
-50 \\
+\ 9 \\
\hline
\cdot \\
\cdot \\
\cdot \\
+\ 4
\end{array}
\\[1em]
q_1 = 3 & q_2 = -6 & q_3 = 5 & q_4 = -6
\end{array}
$$

The pseudo-quotient Q', which has both positive and negative digits, is given by

$$ Q' = .3\ \bar{6}\ 5\ \bar{6}\ 5\ \bar{6} \cdots $$

where the overbars indicate negative digits. After converting the pseudo-quotient to conventional form, we find the quotient Q to be

$$ Q = .2\ 4\ 4\ 4\ 4\ 4\ 4 \cdots $$

as in the restoring method.

Observe, however, that if we have performed $(r - 1)$ subtractions without overdrawing, we simply record the pseudo-quotient digit in that position as $(r - 1)$, shift the remainder, and continue our subtractions in the next position. That is, the pseudo-quotient digits do not necessarily alternate in sign as they did in the previous example. For instance, 910/101 yields a pseudo-quotient Q' given by

$$Q' = 9.1\ \bar{9}\ \bar{1}\ 9\ 1\ \bar{9}\ \bar{1} \cdots$$

from which $Q = 9.0\ 0\ 9\ 9\ 0\ 0\ 9 \cdots$.

An important matter in these successive subtraction algorithms is the question of the initial alignment of the dividend and the divisor. If the divisor is too far to the right of the dividend, a great many subtractions may occur before the remainder goes negative. The divisor should initially be far enough left so that no more than $r - 1$ subtractions will be needed to cause an overdraft. For example, in dividing 2307 by 9, we should not start out

$$
\begin{array}{ccc}
\begin{array}{r} 2307 \\ -\ \ \ \ 9 \\ \hline 2298 \\ -\ \ \ \ 9 \\ \hline \end{array}
& \text{or} &
\begin{array}{r} 2307 \\ -\ \ \ 9 \\ \hline 2217 \\ -\ \ \ 9 \\ \hline \end{array}
\\
\quad\vdots & & \quad\vdots
\end{array}
$$

but rather

$$
\begin{array}{ccc}
\begin{array}{r} 2307 \\ -\ \ 9 \\ \hline 1407 \end{array}
& \text{or} &
\begin{array}{r} 2307 \\ -9\ \ \ \\ \hline -6693 \end{array}
\end{array}
$$

With a numerator of $2n$ digits and a divisor of n digits, we start our algorithm by aligning the divisor beneath the leftmost n digits of the numerator. The set of quotient digits that are formed by our algorithm is independent of the actual position of the radix point in the operands. If our example had been either 2.2/9 or 22/9 or 2200/9, starting the subtraction with the 9 below the second 2 would result in the same pseudo-quotient digits $3\ \bar{6}\ 5\ \bar{6}\ 5\ \bar{6} \cdots$, but the radix point would be in different locations in these three cases. Ultimately, the programmer must decide where to put the radix point; the computer manipulates only the digits.

The usual practice in hardware structures for division is to treat the dividend and the divisor as though their radix points are equally distant from the left, and no prealignment whatsoever is performed. If, under this assumption, the divisor is smaller in magnitude than the dividend, a trap to an overflow procedure occurs and the division is not executed. Otherwise, the division procedure is executed and the quotient will be a fraction. Treating all numbers in our algorithms as fractions is a procedure that is compatible with floating-point arithmetic, in which the significant digits are customarily expressed as a left-justified fraction. In such normalized floating-point arithmetic, any overflow due to the dividend's exceeding the divisor will be at most one position and is easily adjusted by a one-position shift and a corresponding alteration of the exponent.

For the moment, we will assume the dividend magnitude to be smaller than the divisor magnitude. The algorithm for nonrestoring division may be stated as follows:

1. Shift the remainder (initially the dividend) one position left.

2. (a) If the signs of the remainder and the divisor are the same, subtract the divisor repeatedly until the result changes sign or until $(r - 1)$ subtractions have been performed. The pseudo-quotient digit is positive and is equal to the number of subtractions.

   (b) If the signs of the remainder and the divisor are different, add the divisor repeatedly until the result changes sign or until $(r - 1)$ additions have been performed. The pseudo-quotient digit is negative and is equal to the number of additions.

Thus at each step of this algorithm we determine q_j, the next digit of the pseudo-quotient, and x_{j+1}, the next partial remainder, where these values are related by

$$x_{j+1} = r \times x_j - q_j \times D \qquad \text{for } j = 0, 1, 2, \ldots, (n - 1)$$

x_0 is the 0th remainder [i.e., the dividend (numerator) N], D is the divisor, and r is the radix. n iterations of this procedure generate a remainder and a set of pseudo-quotient digits such that

$$\underbrace{x_n \times r^{-n}}_{\substack{\text{Shifted} \\ \text{remainder}}} + D \times \underbrace{\sum_{j=0}^{n-1} q_j \times r^{-(j+1)}}_{\text{Pseudo-quotient}} = x_0$$

That is, the pseudo-quotient $Q' = q_0 \times r^{-1} + q_1 \times r^{-2} + q_2 \times r^{-3} + \cdots q_{n-1} \times r^{-n} = .q_0 q_1 q_2 \ldots q_{n-1}$. Note that since the algorithm can produce both positive and negative pseudo-quotient digits, a radix r system gives rise to digits $\{ -(r - 1), -(r - 2), \ldots, -1, +1, \ldots, +(r - 1)\}$. The digit 0 does not occur as a pseudo-quotient digit.

This algorithm is also valid for numbers expressed in complement form. Consider, for example, 2.2/−9 and replace −9 by its radix complement 991. We now carry out our algorithm, retaining three digits in each operation so that sign information is not lost in the left-shift step:

$$
\begin{array}{rl}
& \text{Signs} \\
& \downarrow \\
r \times x_0 = & 022 \\
D = & \underline{991} \\
& 013 \\
& \underline{991} \\
& 004 \\
& \underline{991} \\
x_1 = & 995
\end{array}
\qquad
\begin{array}{l}
\text{Signs differ—add } D \text{ repeatedly until} \\
\quad \text{result changes sign} \\
\\
\\
\\
\\
\text{Sign has changed after three additions—} \\
\quad q_0 = \bar{3}
\end{array}
$$

$$r \times x_1 = 950 \qquad \text{Shift previous result one position left}$$
$$D = 991 \qquad \text{Signs agree—subtract } D \text{ repeatedly by}$$
$$\qquad\qquad\qquad \text{adding its complement}$$

$$
\begin{array}{r}
950 \\
\underline{009} \\
959 \\
\vdots
\end{array}
$$

$$
\begin{array}{r}
\underline{009} \\
\end{array}
$$

$$x_2 = 004 \qquad \text{Sign has changed after six subtractions—}$$
$$q_1 = 6$$

$$r \times x_2 = 040 \qquad \text{Shift previous result one position left}$$
$$D = \underline{991} \qquad \text{Signs differ—add } D \text{ repeatedly}$$
$$031$$
$$\vdots$$

$$
\begin{array}{r}
\underline{991} \\
995
\end{array}
\qquad \text{Sign has changed after five additions—}
$$
$$q_2 = \bar{5}$$

Continuing, we find

$$Q' = .3\ 6\ \bar{5}\ 6\ \bar{5}\ 6 \cdots$$

which is equivalent to

$$Q = -.2\ 4\ 4\ 4\ 4 \cdots$$

5.3 BINARY NONRESTORING DIVISION

As already noted, the algorithm for nonrestoring division can produce both positive and negative quotient digits, $\pm 1, \pm 2, \ldots, \pm(r - 1)$. For $r \neq 2$, this is a form of redundant representation of numbers, involving $2(r - 1)$ symbols instead of merely r. For the particular case of $r = 2$, $2(r - 1) = 2$ and we may choose to use $+1$ and -1 as the two symbols. We first restate our nonrestoring division algorithm for the binary case, using $+1$ and -1 for the symbols. We will then alter the statement of the algorithm for the more usual 1 and 0 notation, and finally we shall see how to include a preliminary test for overflow as a part of the algorithm.

The algorithm is as follows:

Let

$x_0 = $ double-length numerator

$x_j = j$th remainder

$q_j = j$th pseudo-quotient digit (from the most significant position)

Then

$$x_{j+1} = 2x_j - q_j \times \text{(divisor)}$$

where

$$q_j = \begin{cases} +1 & \text{if the signs of } x_j \text{ and the divisor agree} \\ -1 & \text{if the signs of } x_j \text{ and the divisor differ} \end{cases}$$

As before, we commence the algorithm with the n-bit divisor aligned with the leftmost n bits of the numerator. We regard the quotient as a fraction (for the moment assuming the denominator magnitude larger than the numerator) and proceed step by step to form the pseudo-quotient.

$$Q' = .q_0 \, q_1 q_2 \cdots q_{n-1}$$

where n is the number of divisor digits, and the dividend is taken to be of length $2n$. To convert this group of digits to the conventional form (in terms of 1's and 0's), let a denote the weighted sum of all the positive digits in Q' and b the weighted sum of all the negative digits in Q'. Then Q, the quotient, is given by

$$Q = a + b$$

We observe that

$$a - b = 0.111 \cdots 1 = 1 - 2^{-n}$$

We can express Q in terms of only the positive digits by adding these two expressions to get

$$2a = Q + 1 - 2^{-n}$$

$$Q = 2a - 1 + 2^{-n}$$

That is, to form Q, we ignore the b group (the negative digits), shift a one position left to double it, subtract 1, and add 2^{-n}. Note that subtracting 1 from the doubled a merely changes the sign digit, and adding 2^{-n} simply puts a 1 in the least significant digit position, which had been vacated by the left shift of a. This procedure for recoding the quotient from its form in terms of $+1$'s and -1's is equivalent to replacing each digit group that is of the form $1 \, \bar{1} \bar{1} \cdots \bar{1}$ by $0 \cdots 1$).

Since the negative digits are not involved, the algorithm may be readily restated in terms of 1's and 0's as follows:

1. (a) If the signs of the remainder and the divisor agree, shift the remainder one place left and subtract the divisor. Record $q_j = 1$.
    (b) If the signs of the remainder and divisor differ, shift the remainder one place left and add the divisor. Record $q_j = 0$.

2. Repeat the sequence until $n - 1$ digits of the pseudo-quotient have been obtained. Then shift this group one position left to change the sign digit, and insert a 1 in the least significant position.

The algorithm is still not in final form, since we must consider an initial test for overflows, and we should also examine the special case of occurrence of a zero remainder. The zero remainder situation is easily disposed of, since it implies that subsequent quotient digits are all zeros. It is possible to continue to execute the algorithm, generating these digits one after the other, but since they are already known, it is possible to include circuitry to test for the zero remainder condition and terminate the algorithm at once.

If we continue to generate quotient digits by means of our algorithm, after the occurrence of a zero remainder, two different situations can occur:

1. With a positive divisor and zero remainder, we form $q_j = 1$, since the signs agree. We subtract the divisor, and on the following steps find all succeeding quotient digits to be 0.
2. With a negative divisor and zero remainder, we form $q_j = 0$, since the signs differ. We add the divisor, and on the following steps find all succeeding quotient digits to be 1.

In the first of these cases, no changes need be made, but in the second case, the infinite sequence of 1's, starting at q_{j+1}, is equivalent to $q_j = 1$ followed by all 0's, as in the first case. Thus, to terminate the division when a zero remainder occurs, we should set the quotient bit at that point to 1 and all subsequent bits to 0, and the final insertion of 2^{-n} should not be performed.

How serious a problem is the zero remainder? Probably not very serious at all. The time lost in unnecessarily carrying the division algorithm through to completion is likely to be miniscule, since division is a relatively infrequent operation, and since zero remainders are relatively infrequent among divisions. Random pairs of binary numbers are not likely to yield zero remainders when they are divided, although the probability of zero remainder is higher for short numbers (4 bits, or 8 bits) than for long numbers, for example, 32 bits. Furthermore, there is a practical difficulty in testing for zero remainder, since the size of the remainder becomes one bit shorter at each step. Thus the number of bits to be tested shrinks from $2n$ to n as the algorithm progresses and as the quotient bits fill the right end of the double-length register (as we shall see in Fig. 5.1). The logic circuitry needed for such a test can become very complex, and a better method to terminate such a division is by use of circuitry for shifting over 1's or 0's. We will examine such methods in later pages when we consider speed-up of division.

Next, we incorporate a step to test for overflow. If we assume that the numbers to be divided have their radix points initially aligned, a simple initial subtraction or addition suffices to determine whether the magnitude of the divisor does indeed exceed the magnitude of the dividend. If it does, the quotient will be a fraction. The following tabulation shows the possible initial steps and their outcomes:

Numerator	Denominator	Action	Outcome
+	+	Form $N - D$	Negative result implies no overflow
+	−	Form $N + D$	Negative result implies no overflow
−	+	Form $N + D$	Positive result implies no overflow
−	−	Form $N - D$	Positive result implies no overflow

That is, if the indicated initial action is performed, an overflow is indicated by a result's having the same sign as the numerator.

Placing this test ahead of our division algorithm makes the algorithm start off with an initial value of $N \pm D$. The resulting quotient will therefore be $(N \pm D)/D = N/D \pm 1$. This ± 1 term can be easily accommodated in the formation of Q from the pseudo-quotient Q', as we shall shortly see.

At this stage of the development our algorithm is as follows:

1. If the signs of numerator and denominator agree, subtract the denominator from the numerator. Set $q_s = 1$. (q_s is the pseudo-quotient bit to the left of the radix point.) If the signs disagree, add the numerator and the denominator. Set $q_s = 0$. Overflow is indicated if the sign of the result = the sign of the numerator. If no overflow occurred, the result of this step is the starting value for step 2.

2. (a) If signs of remainder and divisor agree, set $q_j = 1$; shift remainder one place left; subtract divisor.
   (b) If the signs of remainder and divisor differ, set $q_j = 0$; shift remainder one place left; add divisor.

3. Do step 2 repeatedly until either of the following occurs.
   (a) $n - 1$ quotient digits have been formed.
   (b) A zero remainder occurs before $n - 1$ quotient digits are formed. In this case, set $q_j = 1$ at the point of zero remainder, and set all subsequent quotient bits to zero.

4. Form the true quotient from $Q' = .q_0 q_1 q_2 \cdots q_{n-2}$ by: $Q = 2Q' + 2^{-(n-1)}$ except that $2^{-(n-1)}$ should not be added if step 3(b) was done.

This procedure is equivalent to the following sequence of steps, with each q_j being taken as 1 or 0 according as the signs of the remainder and divisor agree or disagree.

$$x_1 = N - (2q_s - 1) \times D \qquad \text{Overflow detection step}$$

$$\left. \begin{array}{l} x_2 = 2x_1 - (2q_0 - 1) \times D \\ x_3 = 2x_2 - (2q_1 - 1) \times D \\ x_4 = 2x_3 - (2q_2 - 1) \times D \end{array} \right\} \text{Quotient digit steps}$$

Expand x_4 for example:

$$x_4 = 2[2x_2 - (2q_1 - 1) \times D] - (2q_2 - 1) \times D$$

$$= 4x_2 - 2(2q_1 - 1) \times D - (2q_2 - 1) \times D$$

$$= 4[(2x_1 - (2q_0 - 1) \times D] - 2(2q_1 - 1) \times D - (2q_2 - 1) \times D$$

$$= 8x_1 - 4(2q_0 - 1) \times D - 2(2q_1 - 1) \times D - (2q_2 - 1) \times D$$

$$= 8N - 8(2q_s - 1) \times D - 4(2q_0 - 1) \times D - 2(2q_1 - 1) \times D - (2q_2 - 1) \times D$$

Thus we may write

$$N = x_4 \times 2^{-3} + 2\underbrace{\left(q_s + \frac{q_0}{2} + \frac{q_1}{4} + \frac{q_2}{8}\right)}_{Q'} \times D - \underbrace{\left(1 + \frac{1}{2} + \frac{1}{4} + \frac{1}{8}\right)}_{2 - 2^{-3}} D$$

$$N = x_4 \times 2^{-3} + [2Q' - (2 - 2^{-3})]D$$

In general

$$N = x_n \times 2^{-(n-1)} + [2Q' - (2 - 2^{-(n-1)})]D$$

That is, the quotient Q is given by

$$Q = 2Q' - (2 - 2^{-(n-1)}) = 2Q' - 2 + 2^{-(n-1)}$$

where Q' is $q_s .q_0 q_1 \cdots q_{n-2}$. q_s is the bit immediately left of the radix point, and since it is doubled in forming Q, it plays no role in the process, being in the 1 case outside the machine number range and in the 0 case without effect. Similarly, the -2 term is outside the machine number range. We conclude that the first $n - 1$ bits of the correct quotient are obtained by doubling the digit group $\cdot q_0 q_1 \cdots q_{n-2}$, except for our special action in the case of an early zero remainder. Note that in this form of the algorithm, which includes the overflow test, we do *not* alter the digit q_0 in obtaining the true quotient Q from the pseudo-quotient Q'.

Two more matters remain to be settled before our algorithm is complete:

1. Under what circumstances should the correction $2^{-(n-1)}$ be inserted?
2. What is the correct form for the remainder?

We may note first that the insertion of $2^{-(n-1)}$ in the process of recoding the pseudo-quotient to conventional form is a valid step only if the quotient terminates after n digits and consists only of 0's beyond that point. In general, we may expect that there will be a nonzero remainder, and continuation of the division beyond that point would thus produce further quotient digits. Forcing a final 1 in the quotient may thus not be the correct action. A better procedure is to carry out one more step of the algorithm and thus to generate that digit explicitly. We shall follow this practice in the subsequent statement of our algorithm.

Various possibilities exist for the proper form of the remainder. One is to simply retain it as it is; another is to carry out yet another step of the algorithm so as to generate a quotient digit that could be used in rounding the retained digits of the quotient; and another is to put it in such a form that if it were used as a numerator, dividing by the same divisor and using the same algorithm would generate further correct digits of the quotient. Any of these procedures can be defended, but the designer should clearly state to the user which one has been chosen.

Adjusting the remainder so as to put it in the correct form for continuing the division may be done after the last step of the division procedure, in the following way:

1. R, the remainder to be adjusted, is formed after the determination of the last quotient digit.
2. If the signs of R and the divisor agree, subtract D from R. If the signs of R and the divisor differ, add D to R.

The resulting remainder may then be used as a new dividend to form further quotient digits. Note that our division algorithm starts by either adding or subtracting D, and in either case that action exactly cancels our adjustment step, again forming the remainder x_{n+1} to enable us to form succeeding quotient bits.

The hardware registers for executing division can be the same as those used in multiplication, as shown in Fig. 5.1. The double-length numerator initially occupies (AC, MQ), and the divisor is in MD. At each step, the divisor is added to or subtracted from AC, the result is placed in AC, and (AC, MQ) is shifted one position left, vacating the rightmost position of MQ, into which we place the next quotient bit. We do not insert q_s into MQ, but insert $q_0 q_1 \cdots q_{n-1}$, where now q_{n-1} is not forced to be 1 but is set by the next test of the signs of the remainder and divisor. After this last

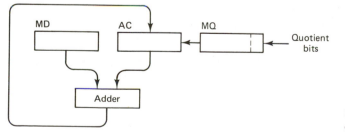

Figure 5.1   Register configuration for division.

step has generated n quotient bits, the final adjustment of the remainder is performed if desired.

The complete algorithm, including remainder adjustment, is given below.

Algorithm 5.1.   Let x_0 = double length (i.e., $2n$-bit) contents of (AC, MQ). Then:

1. Form $x_1 = x_0 + D$ if signs of x_0 and D differ or $x_1 = x_0 - D$ if signs of x_0 and D agree. If signs of x_0 and x_1 agree, then exit to OVERFLOW procedure.
2. Repeat, for $j = 1, 2, \ldots, n$.
   (a) Form $q_{j-1} = 1$ if signs of x_j and D agree or $q_{j-1} = 0$ if signs of x_j and D differ
   (b) Shift (AC, MQ) one position left to form $2x_j$, and insert q_{j-1} into the vacated position at the right.
   (c) Form $x_{j+1} = 2x_j - (2q_{j-1} - 1) \times D$.
3. Adjust the remainder.
   (a) If signs of x_{n+1} and D agree, subtract D
   (b) If signs of x_{n+1} and D differ, add D.

After completion of this algorithm, MQ contains the first n bits of the quotient, and AC contains the remainder in an adjusted form that allows it to be used in turn as a numerator with the same divisor as inputs to this algorithm, so that the resulting quotient can be concatenated with the original one to produce a correct double-length quotient if desired.

It should be noted that this algorithm does not include steps for sensing whether the remainder is zero. As was pointed out earlier, sensing a zero remainder is complicated by the changing size of the remainder at each step. However, if we wish to include this step, we would modify step 2 to appear as follows:

2. Repeat, for $j = 1, 2, \ldots, n$.
   (a) If $x_j = 0$, then shift (AC, MQ) left one position and insert $q_{j-1} = 1$ at the right; shift (AC, MQ) left by $(n - j)$ more positions, inserting 0's at the right; exit from the procedure.
   (b) If $x_j \neq 0$, then follow step 2 as stated originally.

In the absence of the explicit inclusion of this step, the algorithm nevertheless yields a remainder of zero after the adjustment step if a remainder of zero occurred at any point in the procedure.

As an example of the execution of the algorithm, we will follow the contents of AC and MQ through the steps of dividing 0.011011011 by 0.110.

		AC Contents	MQ Contents	
$D = 0.1100$	x_0	0.0 1 1 0	1 1 0 1 1	Initially
		1 0 1 0 0		OF test—subtract D
	x_1	1 1 0 1 0	1 1 0 1 1	Signs differ—$q_0 = 0$
	$2x_1$	1 0 1 0 1	1 0 1 1 0	Shift, and insert q_0
		0 1 1 0 0		Add D
	x_2	0 0 0 0 1	1 0 1 1 0	Signs agree—$q_1 = 1$
	$2x_2$	0 0 0 1 1	0 1 1 0 1	Shift, and insert q_1
		1 0 1 0 0		Subtract D
	x_3	1 0 1 1 1	0 1 1 0 1	Signs differ—$q_2 = 0$
	$2x_3$	0 1 1 1 0	1 1 0 1 0	Shift, and insert q_2
		0 1 1 0 0		Add D
	x_4	1 1 0 1 0	1 1 0 1 0	Signs differ—$q_3 = 0$
	$2x_4$	1 0 1 0 1	1 0 1 0 0	Shift, and insert q_3
		0 1 1 0 0		Add D
	x_5	0 0 0 0 1	1 0 1 0 0	Signs agree—$q_4 = 1$
	$2x_5$	0 0 0 1 1	0 1 0 0 1	Shift, and insert q_4
		1 0 1 0 0		Subtract D
	x_6	1 0 1 1 1		Adjust: Signs differ,
		0 1 1 0 0		so add D
Final:		0 0 0 1 1	0 1 0 0 1	

$$\underbrace{\text{Adjusted}}_{} \qquad \underbrace{\text{Quotient}}_{} = q_0.q_1q_2q_3q_4$$

Adjusted
remainder

However, if the algorithm had been terminated after q_2, we would have used $x_4 = 1101011$ as x_{n+1}. Its sign differs from that of D, and we would adjust it by adding D to form 0011011. To form further quotient digits, that adjusted remainder would be used as x_0 in the first step of a new execution of the algorithm, from which the correct subsequent quotient digits may be readily determined.

An alternative treatment for the remainder R is to adjust it to a value R' such that

$$N = Q \times D + R'$$

To do this, we note that after n quotient digits have been found, we have

$$N = x_{n+1} \times 2^{-n} + [2Q' - (2 - 2^{-n})] \times D$$
$$= x_{n+1} \times 2^{-n} + Q \times D + 2^{-n} \times D$$
$$= (x_{n+1} + D) \times 2^{-n} + Q \times D$$

since, as we saw earlier, the digits of Q are simply those of Q'. Thus we may obtain the modified remainder R' by adding D to the last remainder formed after determination of the nth quotient bit:

$$R' = R + D$$

In our previous example, this action is the same as the adjustment to allow continuing the division, but if we had stopped after $Q = 0.100$, leaving $x_5 = 0.00011$, then we would form $x'_5 = 0.11011$, so that

$$\underbrace{0.011011011}_{N} = \underbrace{0.100}_{Q} \times \underbrace{0.1100}_{D} + \underbrace{0.11011}_{x'_5} \times 2^{-4}$$

5.4 ARRAYS FOR DIVIDING [1–5]

The basic add-and-shift structure of Fig. 5.1 can be replaced by an array of add/subtract elements in much the same way that the multiplication structure was modified in Chapter 4. In division, however, such a structure has no speed advantage, since each addition or subtraction of the divisor must be carried to completion before the correct action for the next row can be determined. The principal advantage of the array is its regular geometric pattern, which lends itself well to incorporation into very large scale integration (VLSI) circuit technology.

The elements that we employ in such an array are adder/subtracter units whose action is determined by an external control signal. Figure 5.2 shows a block symbol for the unit.

A complete array for division of a 7-bit number by a 4-bit number (each either positive or in 2's-complement form) is shown in Fig. 5.3. The control signal introduced at the left of each row is assumed connected to all units in the row and each row also receives the divisor on one of its sets of inputs. In accordance with Algorithm 5.1, we subtract the divisor if the signs of the divisor and the remainder agree, and otherwise we add. This determination is performed by the first four exclusive-NOR gates. The quotient digits are formed by the last four, and the remainder is formed at the bottom. It should be observed that the remainder digits are in an unadjusted form, and another row can be added to the array to add D if the remainder does not have the same sign as the quotient. The reader will find it instructive to trace through the actions of this structure for some typical input set such as $N = 0101101$ and $D = 0110$, for which the structure should give $Q = 0111$ and $R = 0011$.

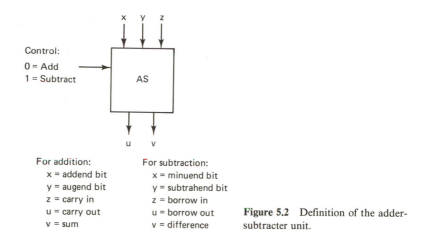

For addition:
    x = addend bit
    y = augend bit
    z = carry in
    u = carry out
    v = sum

For subtraction:
    x = minuend bit
    y = subtrahend bit
    z = borrow in
    u = borrow out
    v = difference

Figure 5.2   Definition of the adder-subtracter unit.

The essentially serial nature of the process is clear, from the fact that the control signal for each row is not determined until the carry/borrow action of the row above is complete. Of course, it is possible to accelerate this action by any of the procedures that we saw in Chapter 3, but these techniques can be quite costly, since they must be used in every row of the structure. It is also clear that this array is amenable to pipelining, if a sequence of divisions is to be performed. A number of variations of the basic structure have been described in the literature.

5.5 SPEEDING UP DIVISION BY SHIFTING OVER 1'S OR 0'S

The division algorithms we have so far examined require either an addition or a subtraction of the divisor in connection with the determination of each quotient bit and are therefore slow algorithms. It is possible to improve on this performance if we assume that numerator and denominator have been normalized (i.e., shifted so that the most significant nonzero digit lies just to the right of the radix point). This is not a serious limitation, since floating-point numbers are usually in that format already, and it is in the floating-point environment that the designer would be most eager to accelerate division.

If the remainder and the normalized divisor are both positive and the remainder has k leading 0's to the right of the radix point, it is clearly smaller than the normalized divisor by a factor of at least 2^{k-1}. Thus the corresponding quotient digits at this point will be a 1 followed by $(k - 1)$ 0's. Our shifting circuitry can be arranged simply to shift the remainder until the next nonzero bit is adjacent to the radix point, while inserting a 1 and $(k - 1)$ 0's at the right end of the quotient register as it is shifted left. Similarly, if the remainder is negative and the divisor is positive, a remainder with k leading 1's to the right of the radix point again represents a quantity smaller in magnitude than the denominator, so that the corresponding quotient bits will be a 0 followed

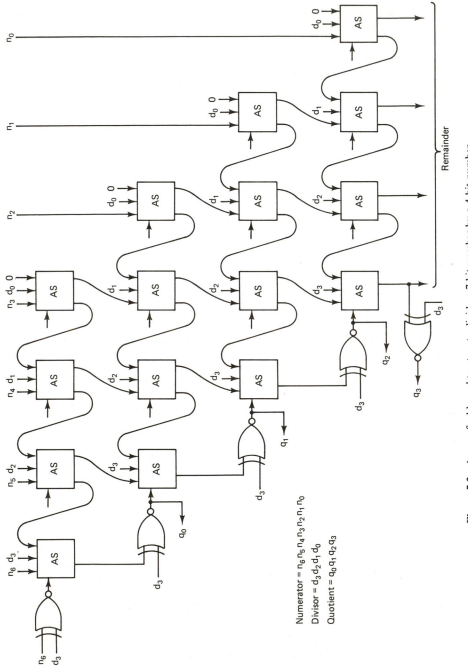

Numerator = $n_6 n_5 n_4 n_3 n_2 n_1 n_0$
Divisor = $d_3 d_2 d_1 d_0$
Quotient = $q_0 q_1 q_2 q_3$

Figure 5.3 Array of adder-subtracters to divide a 7-bit number by a 4-bit number.

by $(k - 1)$ 1's. The remainder is shifted k positions to the left, while the quotient bits are inserted at the right of the quotient register.

We state the algorithm and then give a simple numerical example. Note that our algorithm need not commence with a test for overflow—since we assume normalized numbers, the numerator cannot be more than r times as big as the denominator, and the first trial subtraction will simply show whether the quotient has an integer part or is a fraction. If it has an integer part, we would do a postcorrection consisting of a right shift of the quotient by one position and an increase of the floating-point exponent by one unit. We also assume that our floating point significands are in sign-magnitude form, thus allowing us to treat both numerator and denominator as positive quantities, with the sign to be affixed to the quotient following any required postnormalization.

The algorithm is as follows:

Algorithm 5.2

1. Form $x_1 = x_0 - D$.

2. Repeat, for $j = 1, 2, \ldots, n$.

   (a) If $x_j = \underbrace{0.00 \cdots 01}_{k \text{ 0's}} \cdots$, then form

   $$q_j \, q_{j+1} \cdots q_{j+k-1} = \underbrace{100 \cdots 0}_{(k-1) \text{ 0's}}$$

   and

   $$x_{j+k} = x_j \times 2^k - D$$

   Set $j \leftarrow j + k$.

   (b) If $x_j = \underbrace{1.11 \cdots 10}_{k \text{ 1's}}$, then form

   $$q_j \, q_{j+1} \cdots q_{j+k-1} = \underbrace{011 \cdots 1}_{(k-1) \text{ 1's}}$$

   and

   $$x_{j+k} = x_j \times 2^k + D$$

   Set $j \leftarrow j + k$.

   (c) Otherwise $q_j = \overline{\text{sgn } x_j}$ and $x_{j+1} = 2x_j - (2q_j - 1) \times D$. Set $j \leftarrow j + 1$.

The process may be terminated before $j = n$, if n quotient bits have been determined. We have chosen not to include a final step to adjust the remainder for continuation of the division by another execution of the same algorithm, since it is more common in floating-point machines to provide a separate instruction for handling double-length operands.

As an example, we will divide 0.1110111 by 0.101, first without shifting over 0's and 1's, and then by employing that method.

$x_0 = 01110111$ 00000000

1011   Subtract D

$x_1 = 00100111$ $q_1 = 1$

01001110 00000001   Shift, and insert q_1

1011   Subtract D

$x_2 = 11111110$ $q_2 = 0$

11111100 00000010   Shift, and insert q_2

0101   Add D

$x_3 = 01001100$ $q_3 = 1$

10011000 00000101   Shift, and insert q_3

1011   Subtract D

$x_4 = 01001000$ $q_4 = 1$

10010000 00001011   Shift, and insert q_4

1011   Subtract D

$x_5 = 01000000$ $q_5 = 1$

10000000 00010111   Shift, and insert q_5

1011   Subtract D

$x_6 = 00110000$ $q_6 = 1$

01100000 00101111   Shift, and insert q_6

1011   Subtract D

$x_7 = 00010000$ $q_7 = 1$

00100000 01011111   Shift, and insert q_7

1011   Subtract D

$x_8 = 11010000$ $q_8 = 0$

10100000 10111110   Shift, and insert q_8

We may continue in this fashion, to find $Q = 1.011111001100 \ldots$. (The leading 1 in Q is not a sign digit, of course, but represents a 2^0's component in Q.)

Alternatively, we note that $x_2 = 1.11111100$; a string of six 1's follows the radix point. By our rule, therefore, we have

$$q_2 q_3 q_4 q_5 q_6 q_7 = 011111$$

and

$$x_{2+6} = x_8 = 1.111111 \times 2^6 + 0.101$$

$$= 1.101 \qquad q_8 = 0$$

$$2x_8 = 1.010$$

$$+ D = \underline{0.101}$$

$$x_9 = 1.111 \qquad q_9 = 0$$

and the string of three 1's shows that $q_{10} q_{11} = 11$.

5.6 MORE RAPID DIVISION METHODS

For many purposes, the speed-up achieved by shifting over 1's and 0's in the divisor is quite adequate, especially in view of its modest cost in circuitry. However, as our previous example shows, not all quotient strings of 1's or 0's correspond to remainder strings of 1's or 0's, and we may have to make more uses of the adder than one would infer from counting the number of strings of 1's and 0's in the quotient. In the example, we saw that both q_8 and q_9 were 0, but two adder operations were required to ascertain this fact. Even though x_8 was in normalized form, it was not small enough, relative to D, to allow us to determine more than one quotient digit. The number of quotient 0's and 1's that may be deduced from a given remainder depends on the magnitudes of both the remainder and the divisor.

As we have seen, we always shift the remainder by at least one position in each iteration of the division algorithm, but if we then continue to shift over leading 0's and 1's to normalize the remainder, we may occasionally generate several quotient digits at the cost of only one use of the adder. The average number of shifts per adder use was examined by C. V. Freiman [6], who showed the graph of Fig. 5.4 to exhibit the number of positions of shift involved for different relative magnitudes of normalized remainder and normalized divisor. In each execution of step 2 of Algorithm 5.2, we normalize the remainder by multiplying it by 2^k before adding or subtracting D. The relative magnitudes of the divisor and this normalized remainder determine how many shifts will be performed in the next execution of step 2. Clearly, if they are nearly the same magnitude, they will almost cancel out in the addition/subtraction, and a large number of shifts can occur. This situation corresponds to the region close to the center diagonal of Fig. 5.4.

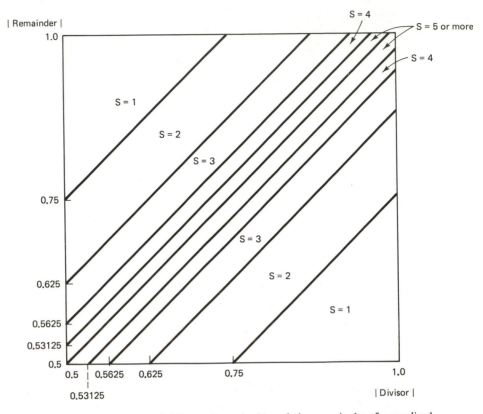

Figure 5.4 Number of shifts, as determined by relative magnitudes of normalized remainder and normalized divisor.

On the other hand, if $D = 0.111$ and the previous normalized remainder is 0.1001, the point corresponding is found in the $S = 1$ region of Fig. 5.4. That is,

$$D = 0.111$$

$$x_j \times 2^k = 0.1001$$

$$x_j \times 2^k - D = 1.1011$$

Single 1 shows that only one shift is peformed at the next step

With $D = 0.111$ and the previous normalized remainder $= 0.1011$, the point is found in the $S = 2$ region. We now find

$$x_j \times 2^k - D = 1.1101$$

Two 1's show that two shifts are performed at the next step

Finally, we let

$$x_j \times 2^k = 0.110101$$

$$-D = \underline{1.001}$$

$$x_j \times 2^k - D = 1.\underline{111101}$$

Four 1's

and four shifts may be executed. The point corresponding may also be found on the graph of Fig. 5.4.

If now one sums the areas of the strips in Fig. 5.4, each being weighted by the corresponding number of shifts, the average value of the shift will be found to be $\frac{8}{3}$, which suggests that our procedure of shifting over leading 0's and 1's will require on the average only $\frac{3}{8}$ as many adder uses as the conventional adder-use-per-bit algorithm.

One method of improving on this shift average is the *divisor multiples method*. At the start, several multiples of the divisor are formed, and at each step that multiple that is closest in magnitude to the remainder is selected. If it were possible to pick that particular multiple which exactly equaled the remainder, all the quotient bits could be found in one step. In practice, of course, only a few multiples can be provided, and we select the one that forms the smallest possible new remainder, so that we may then shift over the largest possible number of leading 1's or 0's. In the IBM 7030 (STRETCH) [7], multiples of $\frac{3}{2} \times D$ and $\frac{3}{4} \times D$ were provided, as well as D itself. A table-lookup procedure, based on the first 5 bits of the divisor and the first 3 bits of the partial remainder, was used to select the appropriate multiple. (Note that the more divisor multiples we use, the more bits of the remainder and the divisor must be examined in making the selection. In the simplest cases, in which only the divisor itself is used, testing the first bit of the positive divisor and the positive remainder suffices.) MacSorley [8] has published results comparing this choice of multiples with other choices and various selection schemes, as well as with simple shifting across 0's or shifting across both 1's and 0's, and finds that use of divisor multiples of $\frac{3}{4}$, 1, and $\frac{3}{2}$ gives shift averages per iteration of about 3.75. With long numbers (as in a scientific floating-point system), this can give substantial time savings as compared with simple nonrestoring division.

5.6.1 SRT Division

The normalizing of remainders characterizes the SRT division method, which received its name from the initials of D. W. Sweeney of IBM, J. E. Robertson of the University of Illinois, and K. D. Tocher of Imperial College, London, who independently and more or less simultaneously devised somewhat similar procedures for division [8–10]. Their methods differ in detail, but share to some degree the following ideas:

1. After each use of the adder, the partial remainder is normalized.
2. Several multiples of the divisor may be provided, one of which is selected at each step.
3. The quotient is represented with a redundant digit set (for example, $\{-1, 0, 1\}$) and requires conversion to conventional form.
4. A higher radix is sometimes used.
5. Through use of redundancy, it is possible to determine quotient digits at each step by inspecting only a few divisor and remainder digits.

The *SRT division method* is in reality not a single algorithm but rather a class of related procedures. These have been developed to a high degree by Robertson and his Illinois colleagues [11–14], whose work forms the basis of the discussion that follows.

In describing the method, we may start from any of the conventional recursion relations for division and display graphically the relation of x_j and x_{j-1} for various allowed quotient digits. For instance, consider

$$x_j = 2x_{j-1} - (2q_j - 1) \times D$$

which is the relation for binary nonrestoring division, with $q_j = \{0, 1\}$. From this we draw Fig. 5.5.

To draw a similar graph for restoring division, recall that here we choose $q_j = 1$ or $q_j = 0$ according as D may be subtracted without overdraft or not. The graph is shown in Fig. 5.6. (Note that in restoring division, only positive remainders are involved.)

Similar graphs may be drawn for division in a higher radix, subject to the recursion relation $x_j = r \times x_{j-1} - q_j \times D$. For example, we may perform nonrestoring division in radix 4, using the digit set $\{-3, -1, +1, +3\}$, and obtain the graph of Fig. 5.7. Quotients produced in this signed-digit form are easily converted to conventional form as they are produced, by replacing each negative digit by its 4's comple-

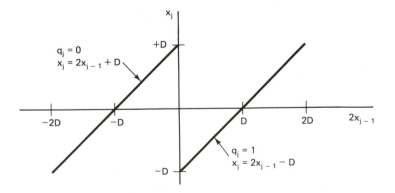

Figure 5.5 x_j and x_{j-1} relations for binary nonrestoring division.

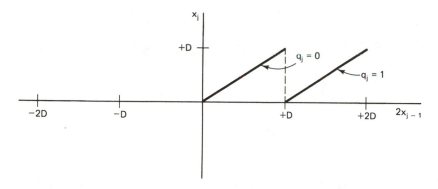

Figure 5.6 x_j and x_{j-1} relations for binary restoring division.

ment and reducing the adjacent higher-order digit by 1. This borrow of 1 cannot propagate, since 0 is not one of the allowed pseudo-quotient digits.

If we now introduce redundancy into the quotient digit set, we have a choice of digits q_j at each value of rx_{j-1}. For instance, in radix 4 division, we might use the digit set $\{-3, -2, -1, 0, 1, 2, 3\}$ to obtain the x_j versus $4x_{j-1}$ graph of Fig. 5.8. For most of the range of values for $4x_{j-1}$, there are two choices for q_j, the better one, of course, being the one that results in the smaller magnitude for x_j.

In Fig. 5.8 we used a set of seven digits, but five could have been used. In general, if we use a digit set $\{-n, \ldots, -2, -1, 0, 1, \ldots, n\}$ having $(2n + 1)$ elements, we must have $2n + 1 \geq r$ so that all r digit values can be represented. However, n, the largest value of any pseudo-quotient digit should not exceed $r - 1$, the largest digit of the radix r set. Thus

$$\left.\begin{array}{r} r > n \\ 2n + 1 \geq r \end{array}\right\} \Rightarrow r > n \geq 1/2(r - 1)$$

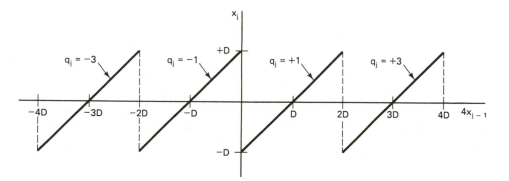

Figure 5.7 x_j and x_{j-1} relations, radix 4 nonrestoring division, digit set $\{-3, -1, +1, +3\}$.

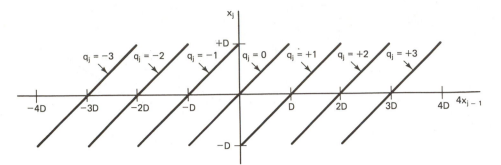

Figure 5.8 x_j and x_{j-1} relations, radix 4 nonrestoring division, redundant digit set $\{-3, -2, -1, 0, 1, 2, 3\}$.

Any convenient value of n in this range may be chosen. For $r = 4$ we might equally reasonably have selected $n = 2$; that is, the digit set could be $\{-2, -1, 0, 1, 2\}$. Now our set of x_j versus rx_{j-1} lines appears as in Fig. 5.9. Since there are only three q_j lines on each side (instead of four as in Fig. 5.8) they are spaced somewhat more widely in order to cover the region defined by ($x_j = \pm D$, $4x_{j-1} = \pm 4D$). Evidently, if $4x_{j-1}$ lies between points 0 and 1, we use $q_j = 0$. Between points 1 and 2, the redundancy allows the choice of either $q_j = 0$ or $q_j = 1$. Over the entire positive region of $4x_{j-1}$, then, our choices are as follows:

Range of $4x_{j-1}$		q_j
0 to 0.5		0
0.5 to 1.0		0 or 1
1.0 to 2.0	$\times D$	1
2 to 2.5		1 or 2
2.5 to 4		2

In comparing these two redundant digit sets for the radix 4 case, we note that the set $\{0, \pm 1, \pm 2, \pm 3\}$ was redundant over almost the entire range of $4x_{j-1}$ ($-3D \le 4x_{j-1} < 3D$); that is, we could make two choices of q_j at most points. On the other hand, to carry out the procedure, we needed to have available not only the divisor but also two times the divisor and three times the divisor. The set $\{0, \pm 1, \pm 2\}$ had less redundancy, but required only D and $2 \times D$. In general, the larger the redundant digit set, the easier becomes the selection process but the more divisor multiples must be made available. With more redundancy (greater overlap of the q_j lines along the abscissa), the precision needed in comparing the divisor and the partial remainder is reduced, so that the time required is less. The price, of course, is the precalculation and the temporary storage of the various divisor multiples.

Now consider binary division using the redundant quotient digit set $\{-1, 0, 1\}$ and with the divisor normalized, that is, $\frac{1}{2} \le |D| < 1$. Our x_j versus $2x_{j-1}$ family appears in Fig. 5.10. In the region of overlap of the $q_j = 0$ line, we choose $q_j = 0$

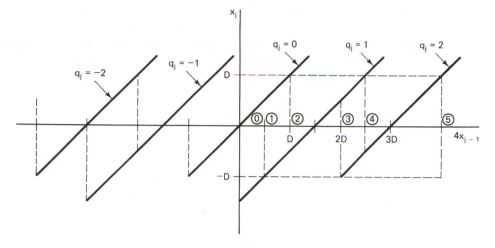

Figure 5.9 x_j and x_{j-1} relations, radix 4 nonrestoring division, redundant digit set $\{-2, -1, 0, 1, 2\}$.

where this gives a smaller remainder than $q_j = 1$ or -1, and otherwise we use the $q_j = 1$ line (or $q_j = -1$ if $2x_{j-1}$ is negative). Since the magnitude of our normalized D cannot be less than $\frac{1}{2}$, we base our selection of $q_j = 0$ or $q_j \neq 0$ on a comparison of $2x_{j-1}$ with the value $\frac{1}{2}$, as follows:

1. If $-\frac{1}{2} \leq 2x_{j-1} < \frac{1}{2}$, then $q_j = 0$.
2. (a) If $D > 0$ and $\frac{1}{2} \leq 2x_{j-1}$, then $q_j = 1$.
   (b) If $D > 0$ and $2x_{j-1} < -\frac{1}{2}$, then $q_j = -1$.
3. (a) If $D < 0$ and $2x_{j-1} < -\frac{1}{2}$, then $q_j = 1$.
   (b) If $D < 0$ and $\frac{1}{2} \leq 2x_{j-1}$, then $q_j = -1$.

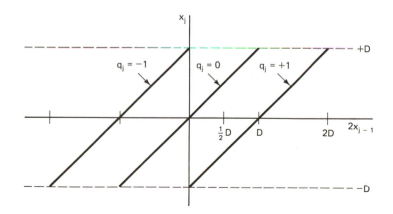

Figure 5.10 x_j and x_{j-1} relations, radix 2 nonrestoring division, redundant digit set $\{-1, 0, 1\}$.

In each case, we then form $x_j = 2x_{j-1} - q_j \times D$. These rules are easy to implement, because the comparison with the value $\frac{1}{2}$ involves only a single bit of precision in the comparison constant. A consequence of this fixed comparison value is that different divisor values will use different portions of the x_j versus $2x_{j-1}$ plot and will be characterized by different values of average shift per iteration. For example, with $D = +\frac{2}{3}$ and partial remainders and the divisor normalized, we find that only a portion of the region in Fig. 5.10 is used. Figure 5.11 shows the family of curves for this case, with the darkened lines indicating those parts of the q_j lines that are used. Figure 5.12 shows the same family with the axes in absolute value, to exhibit the relations more simply.

We may use Fig. 5.12 to determine the probability that q_j is nonzero. Let

$$a = \text{probability that } |2x_{j-1}| \text{ lies between 2/3 and 1}$$

$$b = \text{probability that } |2x_{j-1}| \text{ lies between 1/2 and 2/3}$$

$$c = \text{probability that } |2x_{j-1}| \text{ lies between 1/3 and 1/2}$$

$$d = \text{probability that } |2x_{j-1}| \text{ lies between 0 and 1/3}$$

We note that there are three ranges of $|2x_{j-1}|$ that yield $0 \le |x_j| < 1/6$:

$$0 \le |2x_{j-1}| < 1/6\text{---probability} = d/2$$

$$1/2 \le |2x_{j-1}| < 2/3\text{---probability} = b$$

$$2/3 \le |2x_{j-1}| < 5/6\text{---probability} = a/2$$

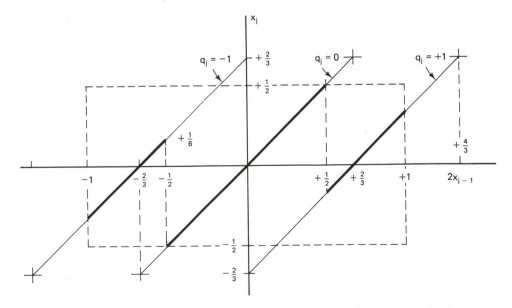

Figure 5.11 x_j and x_{j-1} relations, $q_j = \{-1, 0, 1\}$ and $D = \frac{2}{3}$, showing restricted parts of q_j lines used.

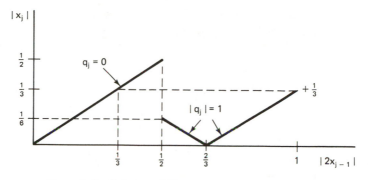

Figure 5.12   Relations of Fig. 5.11 shown in absolute value.

In a steady-state situation:

$$\text{probability}[0 \le |x_j| < 1/6] = \text{probability}[0 \le |2x_{j-1}| < 1/3] = d$$

Thus

$$\frac{d}{2} + b + \frac{a}{2} = d \tag{5.1}$$

Proceeding similarly for $1/6 \le |x_j| < 1/3$ and for $1/3 \le |x_j| < 1/2$, we obtain

$$\frac{d}{2} + \frac{a}{2} = b + c \tag{5.2}$$

and

$$a = c \tag{5.3}$$

We have also

$$a + b + c + d = 1 \tag{5.4}$$

Substituting (5.3) into (5.4) yields

$$2a + b + d = 1$$

Multiplying (5.1) by 2 gives

$$a + 2b - d = 0$$

Thus

$$3a + 3b = 1$$

$$a + b = 1/3$$

That is, the probability is $\frac{1}{3}$ that $|q_j| = 1$, and the average shift in this situation will be three positions.

Other divisor values will yield other probabilities for $|q_j| = 1$, showing that the shift average varies, depending on the divisor value. A more general procedure for calculating the shift average for any divisor value was given by Freiman [6], who showed that a shift average of three positions is obtained over the range

$$3/5 \leq D \leq 3/4$$

A shift average of three positions is of particular interest because of its close correspondence to the canonical form of multiplier recoding, which as we have seen allows a multiplier to be expressed with a minimal number of nonzero digits. With the canonical recoding of the multiplier, we found that on the average one out of three multiplier digits would be nonzero, thus yielding one adder use for each three multiplier digits, on the average.

Recognizing that only a portion of the divisor range gives an average of three positions of shift, Metze [11] proposed a scaling of the divisor to shift it into that range. With divisors in that range, and with a comparison constant $K = \frac{1}{2}$, a quotient range from $K/D_{max} = (\frac{1}{2})/(\frac{3}{4}) = \frac{2}{3}$ to $K/D_{min} = (\frac{1}{2})(\frac{3}{5}) = \frac{5}{6}$ may be defined. This suggests that for divisors outside that range, we may use a different comparison constant K chosen so that

$$6/5\ K \leq |D| \leq 3/2\ K$$

The comparison constant in effect determines the scaling of the division, so that the graph of Fig. 5.12 becomes that of Fig. 5.13. For each divisor, we may find a value of K to scale the division appropriately, and it is clearly advantageous to have only a small number of K values to do this. A given value of K serves for divisors over a range $D_{max}/D_{min} = (\frac{3}{2})/(\frac{6}{5}) = \frac{5}{4}$. We therefore need at least four values of K to cover all possible divisors from $\frac{1}{2}$ to 1 (since $\frac{1}{2} \times \frac{5}{4} \times \frac{5}{4} \times \frac{5}{4} < 1$ but $\frac{1}{2} \times \frac{5}{4} \times \frac{5}{4} \times \frac{5}{4} \times \frac{5}{4} > 1$). Some flexibility exists in choices for K, and Metze has suggested the values in Table 5.1.

The hardware structure for using Metze's method must be able to provide whichever value of K is needed for the particular divisor. The comparison of the remainder and the value of K at each step involves more than one bit (except for $K = \frac{1}{2}$) but need not be executed to full n-bit precision. For these costs, one then obtains a division process yielding an average shift of three positions; that is, the quotient pro-

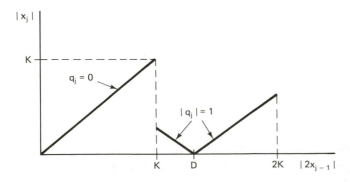

Figure 5.13 x_j and x_{j-1} relations using comparison constant K.

TABLE 5.1

Divisor region	Comparison constant, K
$1/2 \leq \lvert D \rvert < 9/16$	3/8
$9/16 \leq \lvert D \rvert < 5/8$	7/16
$3/5 \leq \lvert D \rvert < 3/4$	1/2
$3/4 \leq \lvert D \rvert < 15/16$	5/8
$15/16 \leq \lvert D \rvert$	3/4

duced will have a minimal number of nonzero digits (although it will not necessarily be in canonical form).

5.7 MORE RAPIDLY CONVERGENT DIVISION PROCEDURES

Computer engineers and computer programmers have always found division to be the slowest and most difficult of the elementary arithmetic operations. Designers of early computers found division to be such a stubbornly intractable problem that they often chose to omit special hardware for division, leaving to the programmer the task of selecting and programming some algorithm whose result would approximate the desired quotient. (As D. R. Hartree remarked in his series of lectures [15] at the University of Illinois in 1948: "Division, which may be rather an untidy process to mechanise, in view of its trial-and-error nature, can be replaced by an iterative process involving only multiplication and subtraction.") This omission of built-in division was justified by the argument that in engineering and scientific computations, it has been found that addition and subtraction operations occur three or four times as frequently as multiplications, which in turn are several times as common as divisions, while in business computations (e.g., payroll, inventory, etc.) divisions are very uncommon. Even today, most microprocessors do not include a division instruction in their command lists and usually omit multiplication, too. However, larger computers capable of stand-alone operation as general-purpose systems usually include division hardware, and for the large scientific computer it is especially important that a fast division structure be provided. This need has become more urgent as fast multiplication schemes have been developed, so as to maintain rough equality between the time spent in multiplications and the time spent in divisions. If K_m and K_d represent, respectively, the number of multiplications and the number of divisions that the machine is expected to execute over a given period, and if T_m and T_d are, respectively, the times needed by the machine to execute a multiplication and to execute a division, then for balance we should have

$$K_m \times T_m = K_d \times T_d$$

$$\frac{T_d}{T_m} = \frac{K_m}{K_d}$$

Thus we seek ways by which division may be speeded up by just as much as multiplication.

A principal difficulty with all the division procedures we have so far examined is that their rate of convergence is only linear; that is, each execution of the iterated procedure adds approximately the same number of bits to the quotient digits already determined. As we have seen, strategems such as the SRT procedure can improve this number, but the procedure is still only linearly convergent, and we seek methods which are more rapidly convergent and not overly costly in hardware.

Some of the more rapidly convergent algorithms have been well known to numerical analysts since the days of the hand-powered desk calculator, where the sheer physical labor of turning the crank was an incentive to improve the speed of the algorithm. The Newton–Raphson method for finding the root of an equation provides the basis for one group of such procedures. Given a function $f(x)$ that is continuous and has a continuous first derivative $f'(x)$, if x_0 is a sufficiently close approximation to a root X of $f(x) = 0$, then a better approximation is

$$x_1 = x_0 - \frac{f(x_0)}{f'(x_0)}$$

This is an iterative procedure, with

$$x_{i+1} = x_i - \frac{f(x_i)}{f'(x_i)}$$

It is easy to show that this procedure converges quadratically. The error ϵ_{i+1} is given by

$$\epsilon_{i+1} = X - x_{i+1}$$

$$= X - x_i + \frac{f(x_i)}{f'(x_i)}$$

$$= \epsilon_i + \frac{f(x_i)}{f'(x_i)}$$

We write a Taylor's series expansion about the point $x = x_i$, limiting the series to three terms:

$$f(X) = f(x_i) + (X - x_i)f'(x_i) + 1/2\,(X - x_i)^2 f''(\xi)$$

where ξ is a value between X and x_i. Since $f(X) = 0$, we have

$$f(x_i) = -\epsilon_i f'(x_i) - 1/2\,\epsilon_i^2 f''(\xi)$$

Thus

$$|\epsilon_{i+1}| = \epsilon_i^2 \times \frac{1}{2} \times \frac{f''(\xi)}{f'(x_i)}$$

and each error is of the order of the square of the error at the previous iteration.

To use this procedure for division, we choose a function that has a root at the desired quotient value or at some value from which we can easily then determine the quotient. To divide N/D, we might select the function

$$f(x) = \frac{1}{x} - D$$

which has a root at $x = 1/D$. Having found this root (to whatever precision we wish), we may then multiply it by N to form the quotient. The graph of this function appears in Fig. 5.14. (D is assumed to be a positive quantity.) It is easy to see that this iterative procedure converges if the first trial value x_0 is chosen so that

$$0 < x_0 < \frac{2}{D}$$

If D is a fraction, then $x_0 = 1$ proves to be a satisfactory starting point for most divisions. This is especially true if D is a normalized fraction, as it would be in floating-point arithmetic.

The iteration rule is

$$x_{i+1} = x_i (2 - D \times x_i)$$

and the error is approximately

$$\epsilon_{i+1} = \epsilon_i^2 \times D$$

This algorithm is one of the best known iterative procedures for division. Its use was considered by Burks, Goldstine, and von Neumann in 1947 [16], who pointed out that a small table of initial estimates for $1/D$ to a precision of 2^{-5} would make it possible to compute $1/D$ to a precision of 2^{-40} in only three iterations. Each iteration requires two multiplications, and the result must then be multiplied by N, for a total of seven multiplications. However, since their multiplier was slow, simple nonrestoring division proved to be faster and was adopted in the design of the Princeton class machines (IAS, ORDVAC, ILLIAC I, and others).

With faster multiplication structures, this iterative algorithm becomes more at-

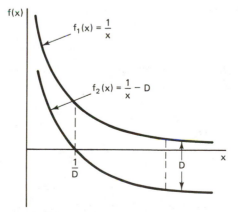

Figure 5.14   Curves of $1/x$ and $1/x - D$.

tractive, and in the 1964 paper [17] in which he introduced the tree of carry-save adders, Wallace described a rapid way of executing the procedure using such a tree.

Higher-order Newton–Raphson formulas can be developed by using more terms of the Taylor's series, so that x_{i+1} is expressed (for example) in terms of both first and second derivatives at $x = x_i$:

$$x_{i+1} = x_i - \frac{f(x_i)}{f'(x_i)} - 1/2 \left[\frac{f(x_i)}{f'(x_i)} \right]^2 \times \frac{f''(x_i)}{f'(x_i)}$$

plus an error term

$$\epsilon_{i+1} = \epsilon_i^3 \times O[f'''(\xi)]$$

This leads to the iterative formula

$$x_{i+1} = x_i[3 \times (1 - D \times x_i) + (D \times x_i)^2]$$

which requires three multiplications per iteration, as well as a "three times" step, which can be done in one addition time. Although it converges more rapidly than the simpler iterative formula, it does not lend itself to a simple implementation in hardware.

Flynn [18] has examined several variants of the Newton–Raphson procedure, but none of them has been found to be distinctly superior to the foregoing ones. Other iterative procedures with good convergence are available, however, and have been used with good results. One of these is a procedure that may be called multiplicative iteration, in which we seek a sequence of factors $m_1, m_2, \ldots, m_n$, such that

$$\lim_{n \to \infty} D \prod_{i=1}^{n} m_i = 1$$

Then we may also form

$$Q \simeq N \prod_{i=1}^{n} m_i$$

where, of course, n is limited to some practical finite value. We derive such a sequence by starting from the function

$$f(x) = \frac{1}{1 + x}$$

which may be expanded by the binomial theorem (if $x < 1$) to give

$$\frac{1}{1 + x} = 1 - x + x^2 - x^3 + \cdots$$

Letting $D = 1 + x$, we have then

$$Q = N \times (1 - x + x^2 - x^3 + \cdots)$$

which may be factored into

$$Q = N(1 - x) \times (1 + x^2) \times (1 + x^4) \times (1 + x^8) \times \cdots$$

Now the 2's complement of a factor $(1 + x^n)$ is $2 - (1 + x^n)$, or $1 - x^n$. We may note then that

$$(1 + x^n) \times (1 - x^n) = 1 - x^{2n}$$

which in turn is the 2's complement of $(1 + x^{2n})$. Thus the successive factors F_{i+1} can be formed by

1. Forming $\overline{F}_i$, the 2's complement of F_i
2. Multiplying $F_i \times \overline{F}_i$
3. Forming the 2's complement of $F_i \times \overline{F}_i$

This procedure converges quadratically, since each pass through it doubles the precision of the result.

This multiplicative iteration algorithm formed the basis of the hardware division structure used in the IBM 360 Model 91, and a good description has been published [19]. It, too, made use of table lookup and a carry-save-adder tree, and was organized to take full advantage of possible concurrencies in executing the algorithm.

REFERENCES

1. Majithia, J. C., Nonrestoring binary division using a cellular array, *Electron. Lett.*, vol. 6, 1970, pp. 303–304.
2. Guild, H. H., Some cellular logic arrays for nonrestoring binary division, *Radio Electron. Eng.*, vol. 39, 1970, pp. 345–348.
3. Stefanelli, R., A suggestion for a high-speed parallel binary divider, *IEEE Trans. Comp.*, vol. C-21, no. 1, Jan. 1972, pp. 42–55.
4. Cappa, M., and V. C. Hamacher, An augmented iterative array for high-speed binary division, *IEEE Trans. Comp.*, vol. C-22, no. 2, Feb. 1973, pp. 172–175.
5. Gardiner, A. B., Comments on "An augmented iterative array for high-speed binary division," *IEEE Trans. Comp.*, vol. C-23, no. 3, Mar. 1974, pp. 326–327.
6. Freiman, C. V., Statistical analysis of certain binary division algorithms, *Proc. IRE*, vol. 49, no. 1, Jan. 1961, pp. 91–103.
7. Buchholz, W., ed., *Planning a Computer System*, McGraw-Hill, New York, 1962, pp. 214–216.
8. MacSorley, O. L., High-speed arithmetic in binary computers, *Proc. IRE*, vol. 49, no. 1, Jan. 1961, pp. 67–91.
9. Robertson, J. E., A new class of digital division methods, *IRE Trans. El. Comp.*, vol. EC-7, no. 3, Sept. 1958, pp. 218–222.
10. Tocher, K. D., Techniques of multiplication and division for automatic binary computers, *Quart. Jl. Mech. Appl. Math.*, vol. 11, pt. 3, 1958, pp. 364–384.
11. Metze, G., A class of binary divisions yielding minimally represented quotients, *IRE Trans. El. Comp.*, vol. EC-11, no. 6, Dec. 1962, pp. 761–764.

12. Atkins, D. E., Higher-radix division using estimates of the divisor and partial remainders, *IEEE Trans. Comp.*, vol. C-17, no. 10, Oct. 1968, pp. 925–934.

13. Atkins, D. E., Design of the arithmetic units of ILLIAC III: use of redundancy and higher radix methods, *IEEE Trans. Comp.*, vol. C-19, no. 8, Aug. 1970, pp. 720–733.

14. Robertson, J. E., The correspondence between methods of digital division and multiplier recoding procedures, *IEEE Trans. Comp.*, vol. C-19, no. 8, Aug. 1970, pp. 692–701.

15. Hartree, D. R., *Calculating Instruments and Machines*, University of Illinois Press, Urbana, 1949, p. 57.

16. Burks, A. W., H. H. Goldstine, and J. von Neumann, Preliminary discussion of the logical design of an electronic computing instrument, Institute for Advanced Study, Princeton, N.J., 1946 (reprinted in C. G. Bell, and A. Newell, *Computer Structures: Readings and Examples*, McGraw-Hill, New York, 1971).

17. Wallace, C. S., A suggestion for a fast multiplier, *IEEE Trans. El. Comp.*, vol. EC-13, no. 1, Feb. 1964, pp. 14–17.

18. Flynn, M. J., On division by functional iteration, *IEEE Trans. Comp.*, vol. C-19, no. 8, Aug. 1970, pp. 702–706.

19. Anderson, S. F., J. G. Earle, R. E. Goldschmidt, and D. M. Powers, The IBM System/360 Model 91: floating-point execution unit, *IBM Jl. R&D*, vol. 11, no. 1, Jan. 1967, pp. 34–53.

EXERCISES

5.1. Carry through the steps of the algorithm for nonrestoring division to find the pseudo-quotient and the true quotient for

(a) $\dfrac{+1000}{+9009}$ operands in sign-magnitude form.

(b) $\dfrac{+73}{+22}$ operands in sign-magnitude form.

(c) $\dfrac{073}{978}$ negative operand in radix complement form.

(d) $\dfrac{+1}{+101}$ operands in sign-magnitude form.

(e) $\dfrac{9035}{1003}$ operands in sign-magnitude form.

5.2. The pseudo-quotient digits generated by the nonrestoring division algorithm depend on the initial alignment of dividend and divisor, but the various pseudo-quotients all convert to the same quotient group in conventional form. Show that division of $482 \div 73$ yields

(a) Pseudo-quotient $= 1\bar{4}.7\overline{9}88\bar{7}$ when the initial alignment is

$$482$$
$$73$$

(b) Pseudo-quotient $= 7.\bar{4}1\overline{8}8\bar{7}$ when the initial alignment is

$$482$$
$$73$$

(c) Pseudo-quotient $= 1\overline{9}4.79\overline{8}8\overline{7}$ when the initial alignment is

$$\frac{482}{73}$$

These three pseudo-quotients are all equivalent to the same true quotient. What is it?

5.3. (a) Show the contents of AC and MQ on successive steps in executing the division of 0.0110110 by 0.110.

    (b) At any point in the process, we should find that $x_0 = Q_j \times D + R'$, where Q_j is given by the first j quotient digits and R' is given by $R' = (x_{j+1} + D) \times 2^{-j}$. Verify this for

$$Q_2 = q_0.q_1$$

$$Q_3 = q_0.q_1q_2$$

$$Q_4 = q_0.q_1q_2q_3$$

5.4. (a) Show the contents of AC and MQ on successive steps in executing the division 1.101101100 $\div$ 0.1010. Adjust the last remainder to put it into proper form for executing the algorithm again to generate further quotient digits.

    (b) Verify that $x_0 = Q_j \times D + R'$ for

$$Q_3 = q_0.q_1q_2$$

$$Q_4 = q_0.q_1q_2q_3$$

$$Q_5 = q_0.q_1q_2q_3q_4$$

Be careful when you multiply the negative Q by the positive D. Booth recoding can be helpful here.

5.5. At any point in the division process, we should find that

$$x_0 = Q'_j \times D + x_{j+1} \times 2^{-j}$$

where $Q'_j = q_0q_1 \cdots q_{j-1}1$, that is, $Q'_j = Q_j + 2^{-j}$.

    (a) Using the numerical examples of Exercises 5.3 and 5.4, verify this relation for

$$Q_2 = q_0.q_1$$

$$Q_3 = q_0.q_1q_2$$

$$Q_4 = q_0.q_1q_2q_3$$

    (b) Show that this procedure is equivalent to forming $Q_j \times D + R'$.

5.6. Using the quotient digit set $\{-3, -1, 1, 3\}$, as shown in Fig. 5.7, carry out the base 4 division of 3302_4 by 13_4. Convert your answer to conventional form (digit set $\{0, 1, 2, 3\}$) and verify it by executing the equivalent base 10 division and converting that result to base 4.

5.7. Compute the weighted average of the strips in Fig. 5.4 and show that the average value of the shift is equal to $^8/_3$.

5.8. Use the iteration rule

$$x_{i+1} = x_i \times (2 - D \times x_i)$$

to find the reciprocal of $0.31830989 \simeq 1/\pi$ to six-decimal-digit precision, starting from $x_0 = 1$. How many iterations are needed? Calculate the difference at each step between x_i and π to observe the behavior of the error.

5.9. Use the multiplicative iteration process to find the reciprocal of $0.31830989 \simeq 1/\pi$ again to six-decimal-digit precision. Compare its convergence with the process of Exercise 5.8.

6
Floating-Point Arithmetic

6.1 ORIGINS

Almost all of the early computers provided only fixed-point arithmetic operations, the only exceptions being the Model V Relay Computer designed by George R. Stibitz of the Bell Telephone Laboratories [1,2] and the Harvard Mark II computer [3] (also largely a relay machine) designed by Howard Aiken. To facilitate scientific and engineering calculations, the early machines often used long word lengths for number representations. Forty-bit numbers were used in machines patterned after the Institute for Advanced Study machine, and 45-bit numbers were used in the SEAC (Bureau of *S*tandards *E*astern *A*utomatic *C*omputer) family. However, merely providing greater precision for number representation did not solve the need for greater range in the size of the numbers. To scale large numbers into the range afforded by the machine involved a great deal of programming effort, as well as a rather thorough analysis of the problem being solved so as to determine the appropriate scaling factors in advance. The technique of automatic scaling, or floating-point arithmetic, came into widespread use in the mid-1950s, first as a software option and then as a hardware feature. Nowadays all computers for scientific and engineering use have built-in floating-point features.

6.2 DEFINITION

A number X may be represented in floating-point form by the triplet (S, E, F), the value of X being taken as

$$X = (-1)^S \times F \times \beta^E$$

146

where S is the sign digit, β is a base value implicit in the floating-point hardware, F is a magnitude (often taken as a fraction), and E is a signed exponent.* Floating-point numbers thus have features in common with numbers in *scientific notation,* in which very large or very small numbers are written without their trailing or leading zeros. We write 3,756,000 as 3.756×10^6, or 0.0003756 as 3.756×10^{-4}, in each case "normalizing" the significant digits by putting the radix point just after the most significant nonzero digit.

In the computer, however, there are much more compelling reasons for using floating-point representations than the simple convenience of suppressing leading or trailing zeros. The principal purpose is to extend the range of numbers representable by digit groups of fixed size. Furthermore, by normalizing the fraction part so that the most significant nonzero digit is at the left, we preserve the maximum precision in representing a number. In fact, since the computer must manipulate numbers in floating-point form and not merely represent them, there has been great interest both from hardware designers and from software engineers in developing consistent procedures for floating-point arithmetic. This interest has been stimulated by advances in VLSI circuit design that are resulting in many new "on-a-chip" computers that have floating-point facilities. The need to ensure program transportability and consistency of results from machine to machine has resulted in the creation of a proposed standard [4] for binary floating-point arithmetic, an effort in which a great many people have been involved.

When floating-point numbers are manipulated in the computer, the group of digits comprising E and F must be accommodated using the fixed-size regions of the storage medium. In virtually all present-day computers, this fixed size is a byte of 8 bits, and a group of such bytes then forms an (S, E, F) triplet. Although the numbers E and F are expressed in a binary form within the computer, they can be in any of several possible binary forms and may even use radices that differ from each other and from the base β. Most frequently, however, the fraction digits

$$f_{-1} f_{-2} \cdots f_{-p}$$

are digits in base β and are regarded as representing the magnitude

$$F = \sum_{-1}^{-p} f_i \beta^i$$

The value of X corresponding to (E, F) is therefore

$$X = (-1)^S \times \beta^E \times \sum_{-1}^{-p} f_i \beta^i$$

If F is normalized, we have $f_{-1} \neq 0$.

---

*These components are somewhat suggestive of logarithms, and it is a common practice to refer to the exponent part as the *characteristic* and the magnitude part as the *mantissa.* We prefer here to be more explicit, using the terms *exponent* and *significand.*

A typical floating-point format is the following one, used in the IBM 370:

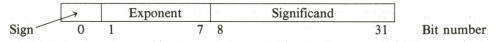

The implied base β is 16. The 2^7 possible exponents correspond to values ranging from -64 to $+63$. The significand magnitude is expressed in 24 bits, which represent six base 16 (i.e., hexadecimal) digits. The magnitude of the largest number that can be represented is therefore

$$\underbrace{.FFFFFF}_{\text{Hexadecimal digits}} \times 16^{+63}$$

which is approximately $7.23 \times 10^{+75}$. The smallest magnitude representable is

$$\underbrace{.100000}_{\text{Hexadecimal digits}} \times 16^{-64}$$

which is approximately 5.4×10^{-79}. In computing this smallest number, we used the smallest fraction expressible in normalized form, that is, with its most significant nonzero digit (in this case, a hexadecimal digit) immediately to the right of the radix point.

(Note that in this format, the significand is given in magnitude form, and the sign bit indicates plus or minus. We could have used a radix complement form for negative significands, but round-off procedures can be performed more consistently with the significand in sign-magnitude form. We shall use this form for all floating-point numbers in the following pages, although, as shown in Table 6.3, this practice is not universally followed.)

It is clear that a set of n digits used for representation of a floating-point number will have greater range but less precision than the same set of digits in a fixed-point representation. To maintain both range and precision, many computers have available a double-length mode, which uses a second n-digit group to extend the significand.

It is also interesting to note that in a *normalized* floating-point format, some of the digit combinations are not used, specifically, those that have leading zeros in the significand. Thus $1/\beta$ of the possible significands are not used. On the other hand, in a nonnormalized format, different representations of the same number are possible if the significand is moved with respect to the radix point. Thus the number of different numbers representable in a floating-point format is somewhat less than if the same bits are used in fixed-point form.

6.3 EXPONENT REPRESENTATION

Before we examine the execution of floating-point operations we should consider the representation of the exponent. Since the exponent must take on both positive and negative values, we have a choice of forms for it, as in the case of fixed-point arithmetic: for example,

1. Sign and magnitude
2. Radix complement or diminished radix complement
3. Biased representation

In the biased representation, a form that offers some useful advantages, a bias quantity equal to half the exponent range is added to each true exponent value to produce a biased exponent (often called the "characteristic"). For example, in the IBM 370 system, 7 bits are available for the exponent, giving 2^7 exponent values. As noted earlier, these values range from -64 to $+63$, and the corresponding binary representations are as shown in Table 6.1. Thus the exponents may be said to be represented in "excess 64" notation.

The principal advantage of the biased form for the exponent arises in the floating-point representation of zero. Since we cannot normalize a number that has no non-zero digits, it is clear that zero does not have a unique representation. Zero in the significand field, together with any value whatsoever in the exponent field, constitutes zero. This can result in some curious ambiguities. For example, if we form

$$0.55 \times 10^{50}$$
$$- \ \underline{0.55 \times 10^{50}}$$
$$0.00 \times 10^{50}$$

and

$$0.55 \times 10^{-50}$$
$$- \ \underline{0.55 \times 10^{-50}}$$
$$0.00 \times 10^{-50}$$

the results are both zero, yet we may legitimately ask whether $.00 \times 10^{50}$ is not really an indication of a quantity much larger than $.00 \times 10^{-50}$, since any round-off that occurred in forming the larger operands was vastly larger than the smaller operands. The IBM 7030 computer (STRETCH) retained such zeros, treating them as "order-of-magnitude zeros," that is, as quantities that were zero to within the order of magnitude of the operands. Present-day practice, however, is to form in such a case a true zero, with the exponent field also set to its lowest value. With the biased form, this lowest exponent value is represented by $00 \cdots 0$, so that a true zero is then represented by $S = E = F = 0$. With either the sign-magnitude form or a complement form for E, a true zero does not have the E digits all zero, a circumstance that can complicate initializing memory cells to zero. A secondary advantage is that by arranging the format as follows:

Sign bit	Biased exponent bits	Significand bits
—	— — · · · — —	— — · · · — —

TABLE 6.1

Biased exponent		True value of exponent (base 10)
Binary	Base 10	
0000000	0	−64
0000001	1	−63
.	.	.
.	.	.
.	.	.
0111111	63	−1
1000000	64	0
.	.	.
.	.	.
1111111	127	63

We find that except for the signs the order of floating-point numbers is the same as the lexicographic order of their representations regarded as binary integers. That is, a set of floating-point numbers of the same sign can be sorted or compared as though they were simply binary integers without any special treatment of their exponent and significand fields.

6.4 PROCEDURES FOR FLOATING-POINT ARITHMETIC

Given operands defined as (E, F) pairs, it is a rather simple matter to define the results of the various arithmetic operations to be executed. To simplify the following discussion, we will not represent signs explicitly, but of course, since each (E, F) pair represents a magnitude, the usual sign procedures for sign-magnitude addition, subtraction, multiplication, and division apply here also. Each factor F_i can be taken in the following as being $(-1)^{S_i} \times F_i$.

Let (E_1, F_1) and (E_2, F_2) be the floating-point representations of X and Y. Then

$$X = \beta^{E_1} \times F_1$$

$$Y = \beta^{E_2} \times F_2$$

$$X + Y = \beta^{E_1} \times (F_1 + F_2 \times \beta^{E_2 - E_1}) \qquad \text{Addition or subtraction}$$

where $E_1 \geq E_2$,

$$X \times Y = \beta^{E_1 + E_2} \times F_1 \times F_2 \qquad \text{Multiplication}$$

$$X \div Y = \beta^{E_1 - E_2} \times \frac{F_1}{F_2} \qquad \text{Division}$$

As a very minimum, therefore, our procedures are as follows:

1. *Addition (or subtraction)*: The number having the smaller exponent must be shifted to the right and its exponent correspondingly increased until the two exponents are the same. The addition (or subtraction) of the resulting significands then takes place.
2. *Multiplication*: The two exponents are added and the significands are multiplied.
3. *Division*: The divisor exponent is subtracted from the dividend exponent to get the exponent of the result. The dividend significand is divided by the divisor significand.

In practice, however, the operations are made more complex by several other factors that must be taken into account:

1. The significand resulting from an arithmetic operation may have a single digit of overflow or may have leading 0's. In either case, a normalizing step must then be performed.
2. When exponents are added in multiplication, the bias is included twice and must be subtracted to form a singly biased exponent. Similarly, in division the biases cancel, and the bias must be added back in again.
3. Arithmetic on exponents can result in either underflow or overflow. These conditions must be recognized and appropriately handled.
4. The result must be in a form that allows any of the possible modes of round-off.

Since round-off plays an important part in the use of any floating-point number system, we consider it in the next few pages before turning to a detailed discussion of the implementation of floating-point arithmetic.

6.4.1 Rounding [5–8]

The rounding off of data is a familiar process in making physical measurements, where we regard the lowest-order significant digit as being correct to within only plus or minus one-half unit in that position. That is, the true value of the quantity being measured is nearer to the (rounded) measured value than to either of the values a unit greater or a unit less than that rounded value in its least significant digit. Most experimenters are aware, too, of the pitfalls involved in performing computations upon such rounded observations.
   Similar problems arise in the computer, but here our problem is not with imprecise measurements but rather with limited digit capacity for representing numbers. The number π, for instance, is known to thousands of significant figures, but we employ only a very few of them in any computation. The need for round-off arises in computation whenever we choose to retain only a part of the digits representing a

number, as, for example, when we keep only n digits of a $2n$ digit product, or when we discard a remainder and adjust the quotient. Facilities for rounding are a usual part of the floating-point instruction repertoire of a computer, but are not ordinarily provided for fixed-point arithmetic. This is so because floating-point arithmetic is used for applications where we wish to preserve the maximum significance in calculations that often involve extended sequences of operations on relatively few data, as in scientific and engineering work. Fixed-point arithmetic suffices, however, for much of the work of commerce and government, and rounding operations are not needed to the degree required in floating point. Machines that have a group of fixed-point decimal operations available do, however, commonly include a base 10 rounding instruction.

When we discard low-order digits, how should we adjust the retained digits (the high-order part)? The most frequently used rounding mode is "round to nearest." The unrounded number, which has more than p digits, is replaced by the nearest p-digit number. If the unrounded number happens to be precisely centered between two adjacent p-digit numbers, we choose the one of them that ends in zero. This choice eliminates bias in the rounding, since sometimes that neighbor is greater than the unrounded number and sometimes less, with approximately equal probabilities in the random case. For example, the following table shows how some six-digit binary numbers are rounded to 4 bits by this rule.

```
1 0 1 0  │  1 0    Centered between 1011 and 1010—choose 1010

1 0 1 0  │  0 1  ⎫
                 ⎪
1 0 1 0  │  0 0  ⎬  Round to 1010
                 ⎪
1 0 0 1  │  1 1  ⎭

1 0 0 1  │  1 0    Centered between 1010 and 1001—choose 1010

1 0 0 1  │  0 1  ⎫
                 ⎪
1 0 0 1  │  0 0  ⎬  Round to 1001
                 ⎪
1 0 0 0  │  1 1  ⎭

1 0 0 0  │  1 0    Centered between 1001 and 1000—choose 1000

1 0 0 0  │  0 1

1 0 0 0  │  0 0
‾‾‾‾‾‾‾
Four-digit high-
order part
```

It is interesting to note that this procedure is consistent with respect to numbers in 2's-complement form, that is, if rounding to nearest increases the magnitude of a positive number, it would also increase the magnitude of the 2's-complement form of that number.

The other rounding modes are the "directed" roundings. We will examine them in Section 6.6.

6.4.2 Structure of the Arithmetic Unit

The structural elements of an arithmetic unit for floating-point addition or subtraction are shown in Fig. 6.1. AC is the accumulator, and EX is a right extension of the accumulator. OSR is an operand storage register. We assume the two significands to have been loaded into AC and OSR, with their respective exponents in E_1 and E_2. The pre-alignment step is performed if $E_1 \neq E_2$, with the significand having the smaller exponent being shifted right and its exponent correspondingly increased until $E_1 = E_2$. Since we have no knowledge of the exponent values until the operands are in place, we cannot tell in advance which should be shifted. We can either provide a right exten-

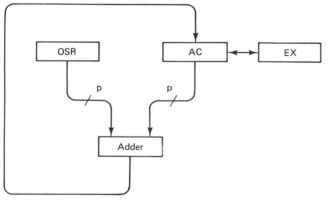

(a)

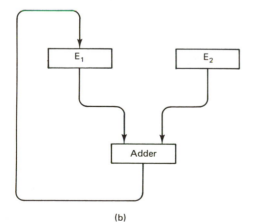

(b)

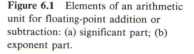

Figure 6.1 Elements of an arithmetic unit for floating-point addition or subtraction: (a) significant part; (b) exponent part.

sion for OSR as well as for AC, which requires extra circuitry, or we can swap the two operands to put the one with smaller exponent into AC and E_1, which requires more time. We assume here the latter of these procedures.

What length should we make register EX, the right extension of AC? It might appear that it should be p digits in length, so that if the addition or subtraction operation forms a result with p leading 0's, we can shift back p digits from EX. In fact, however, this is considerably more than we need for any of the arithmetic operations. The only addition case having more than one leading zero in the result occurs when two nearly equal numbers of opposite sign are added. Two normalized numbers are nearly equal if their exponents are equal and their significands are nearly equal, or if their exponents differ by 1 and the significand of the smaller is almost as big as β times the significand of the larger. For example, the following two situations produce results with several leading 0's:

$$
\begin{array}{lll}
\text{(a)} \quad 0.1111 \times 2^0 & \text{(b)} \quad 0.1000 \times 2^0 & 0.1000 \times 2^0 \\
\underline{-\ 0.1110 \times 2^0} & \underline{-\ 0.1110 \times 2^{-1}} \Big\} \Longrightarrow & \underline{-\ 0.0111 \times 2^0} \\
0.0001 \times 2^0 & & 0.0001 \times 2^0
\end{array}
$$

Since only one digit was shifted into EX [in case (b)], only one digit need be shifted back in during the postnormalization, and it will be followed by 0's. On the other hand, there are many cases in which the result will have a single leading 0, but for these only a single shift is performed in the postnormalization, and again only one digit from EX will be involved, no matter how many prealignment shifts were performed.

A common modern practice is to replace the register EX by three digits G, R, and S, as shown in Fig. 6.2. G corresponds to the leftmost digit of EX, which may have to be shifted into AC in the postnormalization. R corresponds to the next digit of EX and is called the rounding digit because its value determines whether to add a unit into the next-higher-order position in the "round to nearest" mode. The "sticky" bit is merely a record of whether any nonzero digits passed this point in the prealignment step; that is, it "sticks" at 1 if any 1 bit went by. The sticky bit can resolve the question of whether an unrounded number is precisely centered between two p-digit groups or whether it is indeed nearer to one of them.

Thus our register structure of Fig. 6.1 should be modified to appear as in Fig. 6.3. That is, G, R, and S should also participate in the addition or subtraction.

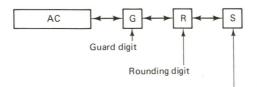

Guard digit

Rounding digit

"Sticky" bit

Figure 6.2 Register EX replaced by G, R, and S.

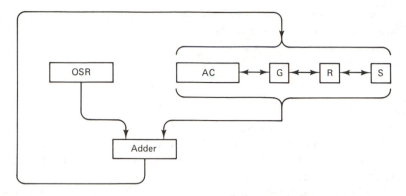

Figure 6.3   Modification of Fig. 6.1 to include G, R, and S.

If the result has leading zeros and must therefore be normalized, the guard digit G is shifted in at the right end of AC on the first left shift and is followed by zeros if more shifts are needed. If more than one left shift is needed, the result has fewer than p significant digits, and may therefore introduce errors into subsequent steps of a computation that treats all operands as having p significant digits. This loss of significance in dealing with numbers of almost the same magnitude is a familiar problem to numerical analysts, who have devised many ways of coping with it (e.g., the rule of thumb that in adding a set of numbers, the small ones should be added first and the larger ones last). Some computer designers have also chosen to permit use of un-normalized numbers, with the least significant digit always at the right and zeros at the left, if there are fewer than p significant digits in a result. We limit our present treatment, however, to the normalized case.

6.4.3 Multiplication

Multiplication of two floating-point numbers starts with the addition of their exponents and the multiplication of their significands and is followed by some simple tests and adjustments of the resulting exponent and significand. Since the significands are fractions, no overflow to left of the radix point can occur, but of course a double-length product can be formed. However, instead of keeping the double-length product, the usual practice is to retain from the low-order part only the G, R, and S positions for use in postnormalization and rounding. A separate set of commands is then provided for doing arithmetic with double-length operands.

When two floating-point numbers are multiplied, the resulting significand may have a leading zero, in which case we must shift left once to normalize. However, the rounding step may then cause an overflow of the significand requiring a right shift to once again normalize. This back-and-forth sequence of shifts could be avoided if, whenever the result of a multiplication has a leading zero, the rounding step is performed in the R position and is followed by a left shift for normalizing if it should be necessary. For instance, consider

$$0 . 1\ 1\ 0 \times 2^0$$
$$\times \underline{0 . 1\ 0\ 1 \times 2^0}$$
$$= 0 . 0\ 1\ 1\ 1\ 1\ 0 \times 2^0$$

A left shift to normalize would form 0.11110×2^{-1} and rounding to nearest would then form 1.000×2^{-1}, which would have to be renormalized to 0.100×2^0. Had we rounded on the R bit, we would at once have formed 0.100 with no need of shifts for normalizing.

Since the exponents are in biased form, the result of the addition of the two exponents contains the bias quantity twice, and the bias must therefore be subtracted to form a correctly biased exponent for the product. The exponent must also be decreased by 1 if the significand is shifted to eliminate a leading zero. This final exponent may be within the allowable range for biased exponent values, or it may be outside this range on either end, producing one of the floating-point singularities known as *exponent overflow* or *exponent underflow*. If the biased exponent is an m-digit binary number $E = d_{m-1}d_{m-2} \cdots d_1d_0$, an overflow or underflow can occur only if d_{m-1} is the same for both operands. The occurrence of (over/under) flow is indicated by $d_m = 1$ in the final exponent (after bias subtraction and significand shift). The nature of the singularity is then indicated by the d_{m-1} bit of either of the original exponents, 1 indicating overflow and 0 indicating underflow.

6.4.4 Division

With minor but obvious modifications, the procedure for division is closely analogous to that for multiplication. The divisor exponent is subtracted from that of the dividend, and the dividend significand is divided by that of the divisor. The resulting significand cannot have a leading zero but can have a digit to left of the radix point, requiring a normalizing shift and addition of 1 to the exponent. The exponent must also be adjusted by addition of the bias quantity, since subtraction of one exponent from the other canceled the bias quantities that both contained. As in multiplication, the final adjusted exponent can have either underflow or overflow singularities.

6.4.5 Handling Exceptional Conditions

The exponent overflow and underflow singularities are two of a group of exceptional situations that may arise in floating-point computations, and we should provide some means of recognizing them as well as some appropriate response. Of course, one method for handling any exceptional condition is simply to set a flag bit in some register accessible to the programmer and then to jump unconditionally from the program to a routine that informs the programmer about the exception. Such "traps" may be necessary in some situations, but they constitute a brute-force way of handling the condition, and techniques are available that in some cases allow the computation to proceed without difficulty and in others at least provide some diagnostic information about the exception.

The most manageable of the exceptional conditions is exponent underflow, a situation that can be handled by relaxing the requirement that all significands be in normalized form. This is a useful alternative to forcing the significand to zero when an exponent underflow occurs. To see how it functions, consider a floating-point number having $\beta = 2$, m exponent bits, and p significand bits. The smallest number representable in normalized form is $\frac{1}{2} \times 2^{-(m-1)} = 2^{-m}$ corresponding to a biased exponent of $00 \cdots 0$. The p significand bits determine $2^{(p-1)}$ equal intervals in the range from 2^{-m} to 2^{-m+1}, and in each succeeding range as the exponent undergoes unit increments, as shown in Fig. 6.4a. If we require all numbers to be either normalized or true zero, the next representable number less than 2^{-m} is zero itself, and no numbers within the region marked "gap" can occur. However, if the biased exponent is held at zero and leading zeros are allowed in the significand, the gap region is then subdivided into $2^{(p-1)}$ intervals of the same size as those between 2^{-m} and 2^{-m+1}, as in Fig. 6.4b. Such "denormalized" numbers reduce the error involved in exponent underflow, making it commensurate with round-off error.

Exponent overflow indicates a number larger than any representable in the system, which may be treated as positive or negative ∞. Some floating-point implementa-

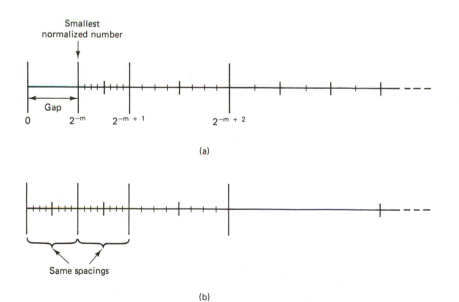

(a)

(b)

Figure 6.4   (a) Representable numbers using normalized format. (b) Representable numbers using "denormalized" format.

tions respond to this situation by returning the machine to its state at the beginning of the instruction and signaling the programmer. The IEEE Proposed Standard [4] sets the exponent field to its maximum value (all 1's, in the binary case) and the significand field to zero. Subsequent operations that manipulate this quantity are then required to give the same results as if the value ∞ was used.

Another important exception is divide-by-zero, a situation that is a result more often of characteristics of the computed data than of carelessness on the part of the programmer. Since division by zero would produce ∞, the processing options for this exception are essentially the same as for exponent overflow, with, however, a unique flag set to indicate the cause of the difficulty.

The IEEE Proposed Standard also includes a class of exceptional results known as "Not a Number"—NaN. These can arise whenever an operation has no mathematical interpretation, as in division of zero by zero. NaNs are identified by an E field which is all 1's (in binary) and an F field which is nonzero.

6.5 REPRESENTATIONAL ERRORS

As a consequence of the exponential factor, the "quantum" size, that is, the interval between one floating-point number and the next, is not the same over the whole range of representable numbers. For any particular exponent value, the various fractions subdivide the range uniformly, but as the exponent increases by one unit, the range subdivisions defined by the fraction increase by a factor of β, as Fig. 6.3b shows in the case of $\beta = 2$. This behavior, of course, is precisely what one expects from a representation scheme in which the same number of significant digits are used for numbers having various exponents.

How well does such a structure represent numbers? That is, since we cannot represent most numbers exactly and must accept instead an approximation that is one of the set of machine representations, what is the representational error, both in relative terms and in absolute terms?

To answer this question we should know something about the way in which numbers are distributed. A strange phenomenon of numbers is that although we may consider them as being uniformly distributed along the infinite number line representing the reals, the leading nonzero digits of numbers are *not* uniformly distributed. That is, in any typical large collection of data, there will be found many more numbers having leading digits of 1 or 2 than there are with leading digits of 8 or 9. This strange circumstance has been entertainingly described by Warren Weaver, in his book *Lady Luck* [9].* Weaver says: "I have been told that an engineer at the General Electric Company, some . . . years . . . ago, was walking back to his office with a book containing a large table of logarithms. He was holding it at his side, spine down; and as he glanced down at the edges of the pages, he noticed that the book was dirtiest at the opening pages and became progressively cleaner—just as though the early parts of the

*"LADY LUCK by Warren Weaver, 1982 Dover Publications, Inc., N.Y., by permission."

book had been consulted a lot, the middle less, and the concluding part least of all. 'But that,' he must have thought, 'is ridiculous. That implies that people most frequently look up the logarithms of numbers beginning with the digit 1, next most frequently numbers beginning with 2, and so on, and least frequently numbers beginning with 9. And this just can't be so; because people look up the logarithms of all sorts of numbers, so that the various digits ought to be equally well represented.'" But as Weaver demonstrates: "The proportion of numbers beginning with n or less is not $n/9$, but is approximately $\log_{10}(n + 1)$. Thus the proportion beginning with 4 or less is $\log_{10} 5 = 0.699$." He cites as an example "that approximately 0.7 of the physical constants in the Chemical Rubber Company tables begin with 4 or less, as compared with the theoretical value 0.699." A related matter, the distribution of the significands of numbers occurring in floating-point computations, has been examined by R. W. Hamming [10], who has shown that "from any reasonable distributions, repeated multiplications and/or divisions rapidly move the distributions toward the reciprocal distribution." He demonstrates that the distributions of the significands of floating-point numbers approach the form

$$r(X) = \frac{1}{X \ln \beta}$$

where X is the normalized significand, β is the base of the number system, and $r(X)$ is the density of the distribution. That is, the number of numbers of a given size is greater for small numbers than for large. The integral of this density function $r(X)$ gives the logarithmic relation cited by Weaver.

The reciprocal distribution of numbers plays a role in the determination of the average error with which numbers are represented in floating-point systems. Let $p =$ the number of binary digits in the fraction part. Then the fraction digits subdivide the range from $1/\beta$ to 1 into intervals of size 2^{-p}. Numbers lying within one of these intervals must be represented by the nearest boundary value. The maximum value of the representation error is thus $\frac{1}{2} \times 2^{-p}$, and the average value of the error in an interval is $\frac{1}{4} \times 2^{-p} = 2^{-(p+2)}$. Let x be the value of the fraction being represented. Then the relative error may be taken as $2^{-(p+2)}/x$. Now if we assume the reciprocal distribution of the fractions, we find more fractions having small values, that is, large relative representation errors. The *average relative representation error* (ARRE) is found from

$$\text{ARRE} = \int_{1/\beta}^{1} \frac{1}{x \ln \beta} \times \frac{2^{-(p+2)}}{x} \, dx$$

$$= \frac{\beta - 1}{\ln \beta} \times 2^{-(p+2)}$$

This formula appears (incorrectly) in a paper by Cody [11], who goes on to compare the ARREs for three floating-point systems having base β, e exponent digits, and p fraction digits. The system and their ARREs are shown in Table 6.2. These three examples were chosen to have the same number range (ratio of largest magnitude repre-

TABLE 6.2

β	e	p	ARRE	MRRE
2	9	22	0.180×2^{-21}	0.5×2^{-21}
4	8	23	0.135×2^{-21}	0.5×2^{-21}
16	7	24	0.169×2^{-21}	2^{-21}

sentable to smallest magnitude representable), which in this case is 2^{2^9}. It may be noted that the ARRE is smallest for $\beta = 4$.

The *maximum relative representation errors* (MRREs) are also tabulated above, MRRE being given by

$$\text{MRRE} = 2^{-(p+1)} \times \beta$$

For this choice of number range, the base $\beta = 4$ turns out to be the best in terms of minimizing representational errors. Of course, its superiority is not dramatic, but it is interesting that it is at least demonstrable.

6.6 *INTERVAL ARITHMETIC AND ITS ROUNDINGS* [12, 13]

Most real numbers cannot be represented exactly within the finite number system of the computer. However, each real number X has two nearest neighbors a and b in the machine number system, where

$$a \leq X \leq b$$

That is, a and b are those members of the set of machine representations that are respectively the greatest lower bound and the least upper bound on X. Since an extended sequence of numerical computations is subject to an accumulation of round-off errors that may be very difficult to analyze, users occasionally resort to calculations based on the intervals in which numbers are known to lie. Thus each number X_j is represented by its interval (a_j, b_j), as defined above.

When two interval numbers X_1 and X_2, are added, the interval in which their sum might be is $(a_1 + a_2, b_1 + b_2)$. We can similarly express intervals for the results of the other elementary arithmetic operations:

Subtraction: $(a_1, b_1) - (a_2, b_2) = (a_1 - a_2, b_1 - b_2)$

Multiplication: $(a_1, b_1) * (a_2, b_2) =$

$$\min (a_1 \times a_2, a_1 \times b_2, b_1 \times a_2, b_1 \times b_2),$$

$$\max (a_1 \times a_2, a_1 \times b_2, b_1 \times a_2, b_1 \times b_2)$$

Division: $(a_1, b_1) \div (a_2, b_2) = (a_1, b_1) \times \left(\dfrac{1}{b_2}, \dfrac{1}{a_2} \right)$

Of course, the new bounds themselves are formed by arithmetic operations in the computer's number system and may be values that are not exactly representable. Thus rounding is needed, in the downward direction for the lower bound and in the upward direction for the upper bound.

In order to support both conventional magnitude rounding and the directed roundings useful for doing interval arithmetic, the following set of rounding operations are included in the IEEE Proposed Standard [4]:

1. *Round to nearest*: Rounded X = the nearer of (a, b) to X. Where a and b are equidistant from X, choose that one whose least significant bit is 0.
2. *Round toward zero*: Rounded X = the smaller in magnitude of (a, b). (This rounding is also known as *chopping* or *truncation*.)
3. *Round toward plus infinity*: Rounded $X = b$.
4. *Round toward minus infinity*: Rounded $X = a$.

The latter two are called *directed roundings*.

We have already examined rounding toward nearest, but some comments should be made about the others. We may note that for numbers expressed in sign-magnitude form, truncating the number by deleting digits at its less significant end always reduces the magnitude of the number (if any nonzero digit is deleted). However, if negative numbers are represented in complement form, truncating them rounds away from zero instead of toward zero. This inconsistency provides a strong incentive for use of the sign-magnitude form in floating-point arithmetic.

The roundings toward plus or minus infinity are needed for the bounds in interval arithmetic. For instance, when we add $X_1 + X_2$ to form $(a_1 + a_2, b_1 + b_2)$, the bound $(a_1 + a_2)$ must be rounded toward $-\infty$, and $(b_1 + b_2)$ must be rounded toward $+\infty$. The sticky bit plays a role here, too, by indicating whether there is any digit of low significance to be taken into account in this rounding step.

Interval arithmetic tends to be overly pessimistic in measuring errors. As the calculation proceeds, the error bounds tend to increase more rapidly than the error itself. Nonetheless, if the calculation is an extremely important one, it may be worthwhile to repeat it, using interval arithmetic and higher precision, to verify the correctness of an algorithm and its result. Obviously, interval arithmetic at least doubles the computation time needed, and carrying out higher-precision operations can add orders of magnitude to the computation time.

6.7 CONVERSION OF BASES AND FORMATS

Conversion of numbers between the decimal form used in the world external to the computer and the floating-point format within the computer is a somewhat more complex problem than the simple radix conversions of Chapter 2 because we must treat both the exponent and the significand of the floating-point number. The conversion would be simplified if the one radix were some integer power k of the other, since a change of k units in the exponent of the smaller radix would correspond to a change of one unit in the exponent of the other, without any effect on the significands. Unfortunately, since we use radix 10 in everyday life, we must cope with it in our conversion process. Since $\log_{10} 2 \simeq .3010300$, a change of 1 unit in the exponent of 10 corresponds approximately to 3.32192810 units in the exponent of 2 (in a binary floating-point number) or 0.83048202 unit in the exponent of 16 (in a base 16 floating-point number).

There are several algorithms from which we may choose in performing the conversion. Consider first the problem of converting a normalized floating-point machine number to a string of binary coded decimal digits together with an indicator of the radix point's position in the result. (This indicator can be a number that tells the position of the radix point, in which case the output has the appearance of a floating-point number, or it can be a properly located radix-point character, giving the output the appearance of a fixed-point number. In either case, however, we must compute its position.)

To simplify our discussion, we will take $\beta = 2$ in the floating-point numbers. The algorithms can be easily modified for other bases. To convert

$$X = F \times 2^E$$

we can start by finding a number $X' = F' \times 10^{E'} = 2^E$, where F' is a normalized binary fraction. For example, if $X = 0.11 \times 2^{13}$, then

$$2^{13} = 8192$$

$$= 0.8192 \times 10^4$$

$$= \underbrace{0.11010001 \ldots} \times 10^4$$

$$\text{Normalized binary fraction}$$

We then form

$$X = F \times F' \times 10^E$$

by multiplying $F \times F'$. The resulting fraction may then be converted to decimal form by a suitable algorithm (Chapter 2), and E' can be incorporated in the result to fix the decimal point.

The most direct way of determining the (F', E') pair that defines X' is simply to store the set of values for all the binary exponents E in a table, either in software or hardware. The complete conversion thus requires a table-lookup operation, a multiplication, and the actual binary-to-BCDD conversion.

Another equivalent algorithm starts instead with the exponent E, multiplying it by $\log_{10} 2$ and rounding the result to the nearest integer E''. E'' is then the nearest exponent of 10 such that $10^{E''} \simeq 2^E$. A stored table of values of $10^{E''}$, each stored as a normalized binary fraction and a power of 2 may be used now, and we then divide $F \times 2^E$ by $10^{E''}$ to get a binary number that is then converted in the usual way to

BCDD form. E'' indicates the radix point location in the result. For example, if $X = 0.11 \times 2^{13}$, we find

$$13 \times \log_{10} 2 = 3.91 \ldots$$

$$E'' = 4 \quad \text{(nearest integer)}$$

$$10^4 = 0.1001110001 \times 2^{14} \quad \text{(by table lookup)}$$

Dividing yields

$$\frac{0.11 \times 2^{13}}{0.1001110001 \times 2^{14}} = \frac{0.011}{0.1001110001} = 0.10011101 \ldots$$

Conversion of this fraction (if we carried out the division process far enough) would yield 0.6144, which (together with $E'' = 4$) gives the answer 6144.

Note that we do not actually need to multiply $E \times \log_{10} 2$, since our table of values can just as readily be addressed by the E value itself. Thus this algorithm requires a table-lookup operation, a division, and then the conversion process. The two algorithms are thus about equally fast.

Conversion from fixed-point (or floating-point) decimal to binary floating point is done in a manner inverse to those above. Given $X = F \times 10^E$, with F not necessarily normalized, we first normalize it and adjust E accordingly. (X might be given as a fixed point number, 317.56, for example, and after normalizing we would write it as $F = 0.31756$, $E = 3$). Then we find (perhaps by table lookup) an X such that $X' = 10^E = F' \times 2^{E'}$, where F' is a normalized binary fraction. F itself is converted to a binary fraction by the appropriate algorithm (Chapter 2), following which the converted value is multiplied by F'. The product is the desired binary floating-point significand, and E' is the exponent. (A normalizing adjustment may be necessary at the end.)

Although the mechanics of converting numbers between fixed-point and floating-point forms are straightforward, the consequences of a sequence of conversions can be serious, since there may be accumulations of round-off errors. This error can become particularly troublesome in conversions back and forth between two different floating-point systems, as might be required when several computers are interconnected in a network. When we convert a fixed-point (FXP) number to floating-point (FLP) form, forming n significant digits in base β, we choose for the FLP equivalent the nearest neighbor in S_β^n (the *significance space*, or set of FLP numbers of n digits to base β). When that number in turn is converted to FLP with respect to a new base δ, we choose this time the nearest neighbor in S_δ^m. Conversion of this number back again to base β and then to FXP or directly back to FXP may or may not yield the original number. The effects of these "compound conversions" have been studied by Matula [14], who points out that the only points unaffected by such conversions are the numbers common to both the source and the destination significance spaces. Base β and δ which are powers of 2 have significance spaces with all their points in common, if we use m base δ digits and n base β digits and take $\delta^m/\beta^n = 1$. Other choices for β, δ, m, and n may result in serious errors in compound conversions.

TABLE 6.3

	IBM 370	Cray-1	Burroughs Scientific Processor	U.S. Air Force MIL-STD-1750A	VAX Digital Equipment Co.	IEEE Standard
Base β	16	2	2	2	2	2
Implicit leading significand bit?	No	No	No	Yes	Yes	Yes
Single precision						
Number of exponent bits	7	15	11	8	8	8
Bias	2^6	2^{14}	None	None	2^7	$2^7 - 1$
Number of significant bits	24	48	37^a	$(23 + 1)^c$	$(23 + 1)^c$	$(23 + 1)^c$
Double precision						
Number of exponent bits	7	Not provided in hardware	11	8	8	11
Bias	2^6		None	None	2^7	$2^{10} - 1$
Number of significant bits	56		72^a	$(39 + 1)^c$	$(55 + 1)^{b,c}$	$(52 + 1)^c$

[a] Both the exponent and the significand are in sign-magnitude form. Double precision uses two numbers, X_1 and X_2, each in single-precision form, so that $X = X_1 + X_2$, with $e_2 \leq e_1 - 36$. The fraction part of X_2 is treated as a 36-bit right extension of the fraction of X_1.

[b] Another double-precision mode is available, having 11 exponent bits and $(52 + 1)$ in the significand.

[c] $(p + 1)$ means p bits appear explicitly and the leading 1 bit is implicit.

6.8 EXAMPLES OF FLOATING-POINT FORMATS

To conclude our discussion of floating-point arithmetic, it is interesting to examine a variety of formats that have been used in actual computers or that have been proposed for such use. Formats for several machines are given in Table 6.3. Several features are now apparent. First, there is no universal agreement on any of the format items. Exponents and significands are of various sizes. Some exponents use the biased form, some use 2's complements, and some use sign-magnitude. Significands are represented in both 2's-complement form and sign-magnitude form. The base β is 16, 8, or 2 in various machines. Some formats that employ base 2 suppress the leading 1 of the significand. (Since it is always present, it is redundant and need not be stored. An extra bit is then available to extend either the significand or the exponent.)

Nevertheless, a few observations can be made about the majority choice in each aspect of the floating-point format:

1. Base 2 is the most widely chosen. Although the range is not so large with a small base as with a large base, the precision is greater, since a large base such as 16 can result in leading binary zeros in the normalized fraction.
2. The format consisting of sign bit followed by biased exponent followed by significand magnitude is the most usual one.
3. Most machines provide double-precision as well as single-precision operations, with the most common structure being 64 bits for double precision and 32 bits for single precision. However, there are widely differing practices in the way the second 32-bit group is used in extending the single-precision format.

The design of optimal floating-point number systems for scientific and engineering computations is a continuing and important topic, and further developments may be anticipated.

REFERENCES

1. Williams, S. B., Bell Telephone Laboratories' relay computing system, Proc. Symp. on Large-Scale Computing Machinery, 1947; *Annals*, Comp. Lab., Harvard Univ., vol. 16, pp. 40–68.
2. Alt, F. L., A Bell Telephone Laboratories' computing machine, *Math. Comp.*, vol. 3, no. 21, Jan. 1948, pp. 1–13.
3. Campbell, R. V. D., Mark II calculator, in Ref. 1, pp. 69–79.
4. IEEE Task P 754, A proposed standard for binary floating-point arithmetic, *IEEE Comp.*, vol. 14, no. 3, Mar. 1981, pp. 51–62.
5. Wilkinson, J. H., *Rounding Errors in Algebraic Processes*, Prentice-Hall, Englewood Cliffs, N. J., 1963.

6. Yohe, J. M., Roundings in floating-point arithmetic, *IEEE Trans. Comp.*, vol. C-22, no. 6, June 1973, pp. 577–586.

7. Kulisch, U., Mathematical foundations of computer arithmetic, *IEEE Trans. Comp.*, vol. C-26, no. 7, July 1977, pp. 610–621.

8. Kuck, D. J., D. S. Parker, Jr., and A. H. Sameh, Analysis of rounding methods in floating-point arithmetic, *IEEE Trans. Comp.*, vol. C-26, no. 7, July 1977, pp. 643–650.

9. Weaver, W., *Lady Luck*, Doubleday (Anchor Books), 1963, pp. 270–277.

10. Hamming, R. W., On the distribution of numbers, *Bell Syst. Tech. Jl.*, vol. 48, no. 8, Oct. 1970, pp. 1609–1625.

11. Cody, W. J., Jr., Static and dynamic numerical characteristics of floating-point arithmetic, *IEEE Trans. Comp.*, vol. C-22, no. 6, June 1973, pp. 598–601.

12. Moore, R. E., *Interval Analysis*, Prentice-Hall, Englewood Cliffs, N. J., 1966.

13. Reuter, E. K., et al., Some experiments using interval arithmetic, *Proc. 4th Symp. on Comp. Arith.*, Oct. 1978, IEEE Cat. no. 78CH1412–6c, pp. 75–80.

14. Matula, D. W., A formalization of floating-point numeric base conversion, *IEEE Trans. Comp.*, vol. C-19, no. 8, Aug. 1970, pp. 681–692.

General

Buchholz, W., ed., *Planning a Computer System*, McGraw-Hill, New York 1962, Chap. 8: Floating point operations.

IBM 370 Principles of Operation, GA22-7000-4, IBM Corp., Data Proc. Divn., White Plains, N. Y.

Knuth, D. E., *The Art of Computer Programming*, vol. 2: *Seminumerical Algorithms*, Addison-Wesley, Reading, Mass., 1969, pp. 180–248.

Liddiard, L. A., Required scientific floating point arithmetic, *Proc. 4th Symp. on Comp. Arith.*, Oct. 1978, IEEE Cat. no. 78CH1412-6C, pp. 56–62.

Payne, M., and D. Bhandarkar, VAX floating point: a solid foundation for numerical computation, *Comp. Architecture News*, vol. 8, no. 4, June 15, 1980, pp. 22–23.

EXERCISES

6.1. Perform the following arithmetic operations and then round each result to the precision of the original operands, using each of the four rounding modes (RN, round to nearest; RP, round toward positive infinity; RM, round toward minus infinity; RZ, round toward 0). Significands are in sign magnitude form. Final results should be normalized.

(a) $(0.101 \times 2^0) \times (0.101 \times 2^0)$

(b) $0.1000 \times 2^0 - 0.1111 \times 2^{-1}$

(c) $0.1000 \times 2^0 - 0.111 \times 2^{-4}$

(d) $0.1101 \times 2^0 - 0.1000 \times 2^{-4}$

(e) $0.1110 \times 2^0 + 0.1001 \times 2^{-4}$

(f) $0.1101 \times 2^0 + 0.1000 \times 2^{-4}$

6.2. Another rounding mode that has sometimes been used is known as "forcing a final 1." In this mode the last bit of the p-digit group to be retained is set to 1 regardless of the digits to be discarded. Although it possesses the obvious advantage of simplicity, this mode also has some significant disadvantages. Explore these, with respect to bias, magnitude and direction of error, and effect on positive and negative numbers in sign-magnitude form or in complement form.

6.3. The following 32-bit group can be interpreted as a floating-point number in IBM format, in IEEE format, or in MIL-STD-1750A format:

$$0100\ 0011\ 1110\ 0110\ 0110\ 0111\ 1110\ 0001$$

(a) Find the number that is equivalent under each of these interpretations. In each case give the sign, the significand magnitude (in binary form), the base, the sign of the exponent, and the true magnitude of the exponent (in binary form).

(b) Not all 32-bit groups can be simultaneously valid in all three of these formats. Try your hand at writing down a few more that are.

6.4. The number of numbers representable in a normalized floating-point format having p binary digits in its significand depends on the base β. If $\beta = 2$, the leading significand bit must be 1, and therefore only $2^{(p-1)}$ significand combinations are allowable. If $\beta = 4$, $\frac{3}{4} \times 2^p$ combinations are allowable, and so forth, for other values of β.

(a) If we assume a floating-point format of e exponent bits and p significand bits (where p is a number divisible by 2, 4, 8, and 16), what is the ratio of $N(\beta)/N(2)$ for $\beta = 4, 8$, and 16? $N(\beta)$ is the number of representations possible in base β.

(b) What is the limiting value of this ratio if $\beta = 2^p$ and $p \to \infty$?

6.5. In Section 6.3 we examined the subject of average and maximum relative representation error. Consider now the question of the maximum and minimum relative and absolute intervals between two adjacent normalized FLP numbers. If two such adjacent numbers are denoted x_1 and x_2, we define their absolute interval to be $|x_1| - |x_2|$ and their relative interval to be $(|x_1 - x_2|)/|x_2|$. What are the maximum and minimum values of these quantities for numbers in IBM format?

7

Some Nonconventional Number Systems

7.1 INTRODUCTION

The various algorithms and number systems that we have so far examined for the elementary arithmetic operations have been almost entirely limited to procedures based on purely binary numbers. There is strong motivation for this limitation in the fact that computer hardware technology is simplest and most reliable when it operates in a binary fashion. Any other number representations that involve digit sets other than {0, 1} must ultimately involve some mode of binary representation, and this circumstance therefore affects our choices of representations and algorithms.

A wide variety of nonconventional number systems have been explored by computer engineers. Many of these have proved to be merely intellectual curiosities, but some have moderate practical advantages in certain situations, and we will examine several of them in the following pages.

7.2 MORE BINARY CODES FOR DECIMAL NUMBERS

In Chapter 2 we examined briefly some of the ways in which binary digits can be used to encode decimal numbers. Although those encodings served our purposes for subsequent discussions, they do not by any means exhaust the encodings that have been found useful.

Binary encodings may be devised with one or more of the following objectives:

1. To facilitate decimal arithmetic
2. To represent decimal numbers compactly
3. To provide redundancy as protection against errors

We shall explore some encodings that serve these purposes.

7.2.1 Self-Complementing Codes

The most efficient codes (in terms of number of bits required per decimal digit) are the 4-bit codes. Most of the very large number of such codes are not systematic and have no use for arithmetic, but those that facilitate the frequent operation of complementation deserve our attention. To be useful for 9's complementation, a 4-bit code should have each digit and its 9's complement represented by groups symmetrically displaced from the center of the list of code groups, so that 9's complementation can be performed simply by changing all four binary digits. If we choose 0101 to represent decimal 1, then 1010 represents 8, for instance. Codes having this symmetry property are known as *self-complementing codes*. A particular one of these that has found wide use is the *excess-3 code* (sometimes written XS-3), in which each digit group is regarded as an 8421 encoding of a value equal to 3 plus the digit represented, as shown in Table 7.1. It is not possible for a weighted code to have the symmetry property, but the excess-3 code does have some desirable arithmetic properties. Thus if we treat excess-3 code groups as 8421 numbers, the sum of two code groups equals the code group for the sum to within an additive constant, and similarly for the difference. That is, the sum of two excess-3 groups is an excess-6 group, from which 3 must be subtracted, and the difference of two excess-3 groups is excess-0 and must be increased by 3. (In a

TABLE 7.1

	0000		
	0001		
	0010		
	0011	$\longleftrightarrow$	0
	0100	$\longleftrightarrow$	1
	0101	$\longleftrightarrow$	2
	0110	$\longleftrightarrow$	3
	0111	$\longleftrightarrow$	4
Center	1000	$\longleftrightarrow$	5
	1001	$\longleftrightarrow$	6
	1010	$\longleftrightarrow$	7
	1011	$\longleftrightarrow$	8
	1100	$\longleftrightarrow$	9
	1101		
	1110		
	1111		

Excess-3 code (bracketing values 0 through 9)

TABLE 7.2

0000	$\longleftrightarrow$	0
0001	$\longleftrightarrow$	1
0010	$\longleftrightarrow$	2
0011		
0100	$\longleftrightarrow$	3
0101		
0110		
0111	$\longleftrightarrow$	4
1000	$\longleftrightarrow$	5
1001		
1010		
1011	$\longleftrightarrow$	6
1100		
1101	$\longleftrightarrow$	7
1110	$\longleftrightarrow$	8
1111	$\longleftrightarrow$	9

sequence of add and subtract steps, some of these adjustments would cancel out, and it might therefore be possible to accumulate them and apply one correction at the conclusion.)

Another useful aspect of the excess-3 code is that carry from one decimal digit to the next is signaled by carry out of the leftmost bit position when two excess-3 numbers are added as 8421 groups.

The technique of carry look-ahead, which is so useful in binary addition, can also be applied in decimal addition. A self-complementing code that facilitates carry look-ahead is shown in Table 7.2. In this code the group 1111 represents 9, and since the code is self-complementing, each pair of decimal digits that add to 9 has a 1 bit in each position of one group where the other has a 0. Thus the carry-propagate signal C_P is easily formed (see also Table 3.3). This code is not an additive code, however, (i.e., the sum of two code groups is not the code group for the sum) and needs some form of read-only memory or programmable logic array to form the sum digits and the C_G signal.

7.2.2 Coding Efficiency

Using a 4-bit code to represent a decimal digit leaves six of the 16 combinations unused. Although this simplifies the input/output conversion problem, it can be costly in storage space. We can do better if we encode two-decimal-digit pairs in 7 bits, wasting then only $^{28}/_{128}$ combinations, or still better with three-decimal-digit groups encoded in 10 bits. Of course, one way to perform such an encoding is to execute a decimal-to-binary conversion by one of the algorithms of Chapter 2, but this costs time. Another way is to store a table of 10^3 10-bit groups corresponding to the 10^3 decimal-digit combinations, but this in itself is expensive in storage. Chen and Ho [1] have described a method of using the 12 bits of the three BCDD groups to form a 10-bit encoding by performing only shifts, deletions, and insertions on the 12-bit group, on the basis of tests performed on the leading bits of the three groups. Their method is described below.

Let the three decimal digits be given in 8421 code, with the bits denoted (*abcd*) (*efgh*) (*ijkl*). The leading digit (*a* or *e* or *i*) of a group is 1 if the group represents 8 or 9 but is 0 otherwise. Thus 0.8 of the possible combinations in each digit position represent "small" digits (i.e., 7 or smaller). In a random distribution, we may expect that all three digits will be "small" in $(0.8) \times (0.8) \times (0.8) \times 100 = 51.2\%$ of the cases, exactly one will be large in $3 \times (0.2) \times (0.8) \times (0.8) \times 100 = 38.4\%$ of the cases, exactly two will be large in 9.6% of the cases, and all three will be large in 0.8% of the cases. Using these probabilities, we assign encodings to the various cases by using a Huffman tree, as shown in Fig. 7.1, where each terminal node is marked with the percentage probability of one of the cases. We note that if all digits are small, $a = e = i = 0$, and the Huffman code of 0 that identifies this case needs merely to be supplemented by (*bcd*) (*fgh*) (*jkl*) to identify the three digits. The codes 100, 101, and 110 which denote exactly one digit large, are supplemented by (*d*) (*fgh*) (*jkl*) if the first number is the large one, by (*bcd*) (*h*) (*jkl*) if the second is large, and by (*bcd*) (*fgh*) (*l*) if the third one is large. We may continue in this way to form 10-bit encodings for all the cases, obtaining the table derived by Chen and Ho and shown as our Table 7.3. To facilitate the translation process, the 5-bit Huffman groups have been separated into two fields, and some of the digit patterns have been put in an order that also simplifies the translation. As with most Huffman encodings, the encodings here are not unique, but they possess the virtue of simplicity, since no arithmetic operations are involved. It is also a straightforward matter to derive the decoding relations that correspond to this table. We may observe, however, that although the encoding and decoding proce-

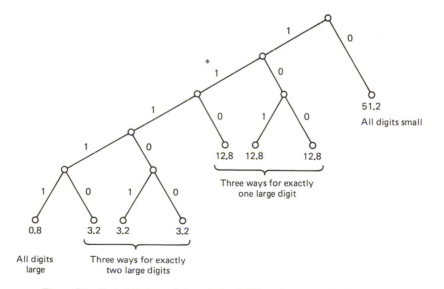

Figure 7.1   Probable sizes of three decimal digits, shown on a Huffman tree.

TABLE 7.3

a	e	i	10-bit encoding									
0	0	0	0	b	c	d	f	g	h	j	k	l
0	0	1	1	0	0	d	f	g	h	j	k	l
0	1	0	1	0	1	d	b	c	h	j	k	l
0	1	1	1	1	0	d	f	g	h	b	c	l
1	0	0	1	1	1	d	0	0	h	b	c	l
1	0	1	1	1	1	d	0	1	h	f	g	l
1	1	0	1	1	1	d	1	0	h	j	k	l
1	1	1	1	1	1	d	1	1	h	0	0	l

dures are reasonable and the representational efficiency is high, the code does not have any simple arithmetic procedures. Therefore, it is useful only for storage, and all numbers passed back and forth between storage and the arithmetic-logical unit must pass through a decoder–encoder structure to be transformed appropriately. The compacting of decimal data in this manner has thus not received much interest from computer designers.

7.2.3 Two Redundant Encodings for Decimal Digits

Two simple encodings for decimal digits that have often been used for detection of single bit errors are the *2-out-of-5 code* and the *bi-quinary code*.

　　The 2-out-of-5 code uses five binary digits to represent a decimal digit, the digit combinations selected being those 10 in which exactly two 1's appear (or those 10 in which exactly two 0's appear). Clearly, any single error changes a 0 to a 1 or a 1 to a 0, forming in either case a group that no longer has two 1's and is therefore recognizable as wrong. Some double errors (but not all) can also be detected, but this encoding is intended primarily for single-error detection.

　　Assignments of code groups to decimal digits can be made in 10! ways with this code, but the usual pattern is the one shown in Table 7.4. This pattern appears some-

TABLE 7.4
A 2-OUT-OF-5 CODE

0	⟷	01100
1	⟷	10001
2	⟷	10010
3	⟷	00011
4	⟷	10100
5	⟷	00101
6	⟷	00110
7	⟷	11000
8	⟷	01001
9	⟷	01010

TABLE 7.5
A BI-QUINARY CODE

0	⟷	01	00001
1	⟷	01	00010
2	⟷	01	00100
3	⟷	01	01000
4	⟷	01	10000
5	⟷	10	00001
6	⟷	10	00010
7	⟷	10	00100
8	⟷	10	01000
9	⟷	10	10000

what arbitrary, but the rightmost four digits of all groups except 0 may be recognized as decimal-digit encodings in the 7421 code. The leftmost bit is then a parity bit chosen to make the total number of 1's in the 5-bit group be an even number. The only 2-out-of-5 pattern not used by the resulting encodings for 1 through 9 is the group 01100, which is assigned to 0. Arithmetic with this pattern of encodings is not quite so simple as with the 8421 code, but is certainly far easier than with a randomly chosen assignment of the 10 groups.

The bi-quinary code uses a base 2 symbol and a base 5 symbol to represent a decimal digit, the base 2 symbol being represented in a 1-out-of-2 code and the base 5 in a 1-out-of-5, as shown in Table 7.5. (Clearly, other assignments of these groups or of the complements of these groups are possible.) This code is also seen to be one that allows detection of all single errors, but its use of 7 bits to do the work of 4 has made it unattractive to computer designers. It was widely used, however, in the days of relay switching networks.

7.3 MIXED-BASE NUMBER SYSTEMS

Mixed-base systems are of interest to the computer engineer not so much because of any intrinsic computational merit that they have (and generally, they haven't any) as because of the circumstance that society employs many mixed-base systems in measuring mass, length, time, money, and other quantities. Some examples of mixed-base systems are shown in Table 7.6.

A special-purpose computer might, with advantage, execute arithmetic directly in some mixed-base system, and we will therefore briefly examine the structure of such systems.

A number representation in a mixed-base system consists of an ordered set of digits $d_n d_{n-1} \cdots d_2 d_1 d_0$, where each digit position j has an associated base b_j and each digit d_j is chosen from among a set of digits allowed in position j, that is,

$$d_j = \{0, 1, \cdots, (b_j - 1)\}$$

TABLE 7.6

Time units:	Week	Day	Hour	Minute	Second	
Base:	—	7	24	60	60	
Distance units: (English system)	League	Mile	Furlong	Yard	Foot	Inch
Base:	—	3	8	220	3	12
Money units: (former British system)	Pound	Crown	Shilling	Pence	Farthing	
Base:	—	4	5	12	4	
Weight units:	Ton	Pound (avoirdupois)	Ounce	Grain		
Base:	—	2000	16	437.5		

The number N of representations that are possible is

$$N = \prod_{j=0}^{n} b_j$$

and if we use the system to represent integers, then the number X represented by $d_n d_{n-1} \cdots d_2 d_1 d_0$ is

$$X = d_0 + \sum_{j=1}^{n} \left(d_j \times \prod_{i=0}^{j-1} b_i \right)$$

(Obviously, fixed-base systems, in which all the bases are the same, are a particular instance of this system, and give rise to $N = b^{n+1}$ and $X = \sum_{0}^{n} d_j \times b^j$.)

We further define a successor relationship in each digit position, so that

Successor of 0   is 1

Successor of 1   is 2

$$\vdots$$

Successor of b_{j-1} is 0

Then, in order to find the representation of the successor of some integer $X = d_n d_{n-1} \cdots d_2 d_1 d_0$, we proceed as follows:

1. $j \leftarrow 0$.
2. $d_j \leftarrow$ successor of d_j.
3. If $d_j = 0$, set $j \leftarrow j + 1$ and go to 2. Otherwise, exit.

This, of course, is simply our usual procedure for counting, which formed the basis of Chapter 3.

Mixed-base systems can of course represent fractions as well as integers. We merely extend our digit group to the right of the "radix point" in the usual way:

$$d_n d_{n-1} \cdots d_2 d_1 d_0 . d_{-1} d_{-2} \cdots$$

and we assign bases to the negative-subscripted positions. We have then

j: $\cdots$	2	1	0	-1	-2	$-3 \cdots$
b_j:	b_2	b_1	b_0	b_{-1}	b_{-2}	b_{-3}
Weight:	$b_1 b_0$	b_0	1	$\dfrac{1}{b_{-1}}$	$\dfrac{1}{b_{-2} b_{-1}}$	$\dfrac{1}{b_{-3} b_{-2} b_{-1}}$
Digit set:	0	0	0	0	0	0
	$\vdots$	$\vdots$	$\vdots$	$\vdots$	$\vdots$	$\vdots$
	$(b_2 - 1)$	$(b_1 - 1)$	$(b_0 - 1)$	$(b_{-1} - 1)$	$(b_{-2} - 1)$	$(b_{-3} - 1)$

It may be convenient (but not necessary) to let the bases for the fraction part be the same as those equally far to the left of the radix point, so that $b_{-1} = b_0$, $b_{-2} = b_1$, $\cdots$, $b_{-j} = b_{j-1}$. The weights of the fraction positions then become

j:	-1	-2	-3	$\cdots$
Weights:	$\dfrac{1}{b_0}$	$\dfrac{1}{b_1 b_0}$	$\dfrac{1}{b_2 b_1 b_0}$	$\cdots$

In representing the digits of a mixed-base system by binary symbols, we usually pay a penalty of some loss of representational efficiency. Since the bases are usually not powers of 2, some of the binary combinations will be unused. For units of distance in the English system, as an example, we need 4 bits to encode the number of inches, 2 for the number of feet, 8 for yards, 3 for furlongs, and 2 for miles.

Arithmetic operations in a mixed-base system are in principle straightforward but must involve addition and multiplication rules for each of the various bases. The rules for addition are a simple extension of those for fixed base—we add the digits of the two operands position by position, and if the sum in any position exceeds b_j for that position, we subtract out b_j and add a carry into the next position (position $j +$ 1). The actual hardware structure for adding would differ from position to position, because of the differences in the bases.

Multiplication is somewhat more complicated. When we multiply two digits x_j and y_j in the jth position of two operands, we are in fact multiplying $(x_j \times w_j) \times (y_j \times w_j)$, where w_j is the weight associated with position j. Our result should be $x_j \times y_j \times w_j$, which is the number of units of weight w_j that should be included in the product. Thus the multiplication table for the jth position is a function not only of digits x_j

and y_j but also of the weight w_j. Consider a mixed-base system involving bases 10, 7, 5, and 3. The associated position weights are:

$$
\begin{array}{lcccc}
j: & 3 & 2 & 1 & 0 \\
\text{Base:} & 10 & 7 & 5 & 3 \\
\text{Weights:} & 105 & 15 & 3 & 1
\end{array}
$$

and the first several positive integers are represented in Table 7.7. To multiply 3×4, form

$$
\begin{array}{r}
0010 \\
\times\ 0011 \\
\hline
10 \\
3 \\
\hline
0040
\end{array}
$$

Consider the multiplication of two mixed-base numbers X and Y, where

$$X = x_3 b_3 b_2 b_1 + x_2 b_2 b_1 + x_1 b_1 + x_0$$

$$Y = y_3 b_3 b_2 b_1 + y_2 b_2 b_1 + y_1 b_1 + y_0$$

where the bases are b_j. We form the product by executing the following multiplications and columnwise additions:

Factors	Number of units having weight $b_3 b_2 b_1$	Number of units having weight $b_2 b_1$	Number of units having weight b_1	Number of units having weight 1
$y_0 \times X$	$y_0 x_3$	$y_0 x_2$	$y_0 x_1$	$y_0 x_0$
$y_1 b_1 \times X$	$y_1 x_3 b_1$	$y_1 x_2 b_1$	$y_1 x_0$	
			$y_1 x_1 b_1$	
$y_2 b_2 b_1 \times X$	$y_2 x_3 b_2 b_1$	$y_2 x_0$		
		$y_2 x_1 b_1$		
		$y_2 x_2 b_2 b_1$		
$y_3 b_3 b_2 b_1 \times X$	$y_3 x_0$			
	$y_3 x_1 b_1$			
	$y_3 x_2 b_2 b_1$			
	$y_3 x_3 b_3 b_2 b_1$			

The sum in each column is formed using the base for that column, with carries from column j being absorbed in column $(j + 1)$.

TABLE 7.7

Integer	d_3	d_2	d_1	d_0
0	0	0	0	0
1	0	0	0	1
2	0	0	0	2
3	0	0	1	0
4	0	0	1	1
5	0	0	1	2
6	0	0	2	0
7	0	0	2	1
8	0	0	2	2
9	0	0	3	0
10	0	0	3	1
11	0	0	3	2
12	0	0	4	0
13	0	0	4	1
14	0	0	4	2
15	0	1	0	0
16	0	1	0	1
17	0	1	0	2

For example, if we use bases 10, 7, 5, and 3, then $23_{10} = (0122)$ and $34_{10} = (0211)$ in this system. We form their product as follows:

$$
\begin{array}{lcccc}
X = & 0 & 1 & 2 & 2 \\
Y = \times & 0 & 2 & 1 & 1 \\
\hline
y_0 \times X: & & 1 & 2 & 2 \\
y_1 b_1 \times X: & & 1 \times 1 \times 3 & 1 \times 2 & \\
& & & 1 \times 2 \times 3 & \\
y_2 b_2 b_1 \times X: & & 2 \times 2 & & \\
& & 2 \times 2 \times 3 & & \\
& & 2 \times 1 \times 3 \times 5 & & \\
\hline
& & 2 & & \\
& 7 & 1 & 0 & 2 \\
\text{product} = & 7 & 3 & 0 & 2 \\
& = & 782_{10} & & \\
\end{array}
$$

Clearly, mixed bases are a mixed blessing, and they make the advantages of the fixed-base system all the more evident.

The Roman numeral system is an interesting example of a structure that possesses some of the features of a mixed-base system. Although by modern standards it may seem clumsy, we should remember that for many centuries it was widely used in the Western world. Skilled practitioners could easily carry out all the computations needed for the daily affairs of commerce and government, and Roman numerals were still used in Europe in the early sixteenth century. Although there were many variations of the notation, for our purposes it suffices to consider the simplest form. The Roman system is, in fact, a decimal system in which we use for each decimal position a base 2 symbol and a base 5 symbol. It is not strictly a positional system, however, since a different group of symbols is used in each decimal position. The relations of the symbols and their positions and weights are shown in Table 7.8. Since each position has its own distinct set of symbols there is no need for the symbol 0. Note that the weight associated with each symbol consists of the product of all the bases of lower-order symbols, just as in the usual mixed-radix system. (This same concept of representing decimal digits by five elements of unit weight and one—sometimes two—elements of weight 5 is seen also in the oriental abacus, the Chinese version having two beads of weight 5, and the Japanese version only one of weight 5.)

Any number (up to a few thousand) can be represented by simply selecting appropriate symbols from this set. However, the notation is made somewhat more compact by the convention that any symbol lying to left of one representing a larger number is subtracted. This convention is unlike any in our modern-day number systems. It has the effect of introducing redundancy, since there are now several ways in which a given number can be represented. For example:

$$IV = IIII$$

$$XXL = XXX$$

$$XIX = XVIV$$

This redundancy was used only to form more compact symbol groups (IV, for instance, instead of IIII), and the modern concept of using redundancy for error protection played no part in the use of Roman numerals.

Another interesting example of the mixed-base form is the *factorial number system*. We again represent a number X by a digit sequence $d_n d_{n-1} \cdots d_2 d_1 d_0$, where the digit set d_j in each position is given by

$$d_j = \{0, 1, \cdots, (j + 1)\}$$

TABLE 7.8

	Thousands		Hundreds		Tens		Units
Roman symbol	M	D	C, CC, CCC, CCCC	L	X, XX, XXX, XXXX	V	I, II, III, IV
Base	5	2	5	2	5	2	5
Weight	1000	500	100	50	10	5	1

and where the numerical value of X is given by

$$X = \sum_{j=0}^{n} d_j \times (j + 1)!$$

The base b_j associated with digit position j is $b_j = j + 2$. The various bases, digit sets, and weights are:

		j:	$\cdots$	3	2	1	0
		b_j:	$\cdots$	5	4	3	2
	Weights:		$\cdots$	24	6	2	1
Digit set:							
0 through			$\cdots$	4	3	2	1

The first few positive integers and their representations are:

X	d_2	d_1	d_0
0	0	0	0
1	0	0	1
2	0	1	0
3	0	1	1
4	0	2	0
5	0	2	1
6	1	0	0
7	1	0	1

7.4 NEGATIVE-BASE ARITHMETIC [2–6]

The use of a negative base is an idea that possesses some superficial appearance of tidiness and compactness, since we do not require a separate bit for representing the sign but instead let the sign be implied by the number itself. -2 is the value most often suggested, and we will examine here only this "negabinary" representation. We consider numbers N of the form

$$N = \sum_{j=0}^{n} d_j \times (-2)^j$$

where $d_j = \{0, 1\}$. (Limiting our discussion to integer values for N simplifies our notation without restricting our generality.) For instance, if we take $n = 3$, the sixteen 4-bit groups have the base 10 equivalents shown in Table 7.9. Note that the 16 groups represent 16 consecutive integers, from -10 through $+5$. Even though there is no sign bit, our 4 bits of course contain no more information than they would in a base $(+2)$ notation.

TABLE 7.9

$(-2)^3$	$(-2)^2$	$(-2)^1$	$(-2)^0$	Base 10 equivalent
0	0	0	0	0
0	0	0	1	1
0	0	1	0	-2
0	0	1	1	-1
0	1	0	0	4
0	1	0	1	5
0	1	1	0	2
0	1	1	1	3
1	0	0	0	-8
1	0	0	1	-7
1	0	1	0	-10
1	0	1	1	-9
1	1	0	0	-4
1	1	0	1	-3
1	1	1	0	-6
1	1	1	1	-5

Absorbing the sign into the number representation (rather than making it explicit) has some interesting consequences for the basic arithmetic operations. We note that numbers in base (-2) can be expressed in terms of the base $(+2)$ as follows:

$$N = \sum_{j=0}^{j=\lfloor n/2 \rfloor} d_{2j} \times 2^{2j} - \sum_{j=0}^{j=\lceil n/2 \rceil - 1} d_{2j+1} \times 2^{2j+1}$$

That is, all even-numbered digit positions make a positive contribution to N and all odd-numbered digit positions make a negative contribution to N (when the digit in that position is 1). For instance, in converting the number 37 from base 10 to base (-2), we proceed as follows:

$$37 \div -2 = -18 + \text{remainder of 1}$$

$$-18 \div -2 = +9 + \text{remainder of 0}$$

$$+9 \div -2 = -4 + \text{remainder of 1}$$

$$-4 \div -2 = +2 + \text{remainder of 0}$$

$$+2 \div -2 = -1 + \text{remainder of 0}$$

$$-1 \div -2 = +1 + \text{remainder of 1}$$

$$1 \div -2 = 0 + \text{remainder of 1}$$

Thus

$$37_{10} = 1100101_{(-2)}$$

$$= 1000101_{(2)} - 0100000_2$$

Note that in our successive divisions by -2, we choose a quotient at each step so as to leave a positive remainder.

Observe also that although n bits still represent exactly 2^n different numbers, the set of numbers represented is asymmetrically distributed about zero. With an odd number of binary digits, there will be more positive representations than negative, and the converse holds for an even number of digits. For large n, this asymmetry is in the ratio 2:1; that is, there will be twice as many representations having one sign as having the other.

Now let us examine the arithmetic operations, starting with shifting. As would be expected, a left shift of k positions multiplies the number by $(-2)^k$, and a right shift of k positions multiplies the number by $[1/(-2)]^k$. The operation of changing the sign of a number, which is so simple in base 2, is somewhat more difficult in base (-2) but can be done by a sequence of shifting and adding. Thus

$$-N = N + (-2) \times N$$

To carry out these steps, however, requires that we determine the rules for addition. Note that

$$0 + 0 = 0$$

$$0 + 1 = 1 + 0 = 1$$

$$1 + 1 = 0, \quad \text{with a carry of 1 into each of the}$$
$$\text{two next more significant positions}$$

Since two 1's in the jth column add up to a value that is the negative of the weight of the $(j + 1)$th column, we will not propagate the carry over the $(j + 1)$th and $(j + 2)$th positions if there is at least one 1 in the $(j + 1)$th position but will instead use the carry to cancel a 1 in the $(j + 1)$th position. Thus our rule should be refined as follows:

c_j	x_j	y_j	s_j	c_{j+1}	c_{j+2}	
0	0	0	0	0	0	
0	0	1	1	0	0	
0	1	0	1	0	0	
0	1	1	0	×	×	"××" = 11 unless x_{j+1} or
1	0	0	1	0	0	$y_{j+1} = 1$. In that case,
1	0	1	0	×	×	$c_{j+1} = -1$ and $c_{j+2} = 0$.
1	1	0	0	×	×	
1	1	1	1	×	×	

For example, to add -9 to $+18$, using the negabinary system, we have

	$(-2)^4$	$(-2)^3$	$(-2)^2$	$(-2)^1$	$(-2)^0$
$-9 =$	0	1	0	1	1
$+18 =$	1	0	1	1	0
	1	1	0	0	1

The two 1's in the $(-2)^1$ column add to -4, which cancels a 1 in the $(-2)^2$ column.

As an example in which a carry must be introduced in the next two positions, consider

$$-9 = 0\ 1\ 0\ 1\ 1$$
$$+14 = 1\ 0\ 0\ 1\ 0$$
$$0\ 0\ 1\ 0\ 1$$

Here the two 1's in the $(-2)^1$ column have produced a carry into the $(-2)^2$ column and into the $(-2)^3$ column. Carries continue to the left, but generate sum digits of zero as far out as we wish to go.

To return briefly to the matter of negating a number, note that our example of $-9 + 18$ accomplished that result, since $+18$ can be regarded as having been formed by shifting -9 one position to the left. The sign reversal is then completed by the addition.

Now that we know the rules for addition and negation, we can undertake multiplication and division by straightforward modification of our algorithms for the radix 2 case, if we wish to do so. However, it is interesting to note that although there exists a moderately extensive literature (several dozen published papers) on the subject of negative-base arithmetic and at least one machine has been built using that system [6], no clear-cut advantages have been found in its favor, in terms of circuit complexity, algorithm complexity, or speed of operation. We will therefore not examine the subject further.

7.5 LOGARITHMS FOR MULTIPLICATION AND DIVISION [7–15]

Ever since the early days of the analog computer, the use of logarithms has attracted the attention of computer engineers as a way of overcoming the difficulties encountered in multiplication and division. However, the advantages of logarithms have proved to be more apparent than real, and practical difficulties associated with generating or storing logarithms render this method impractical except for situations where great precision is not needed. Whatever the method used to form the logarithms and antilogarithms, it must be capable of the precision we need, and it must compare favorably in speed or equipment cost with the algorithmic procedures examined in earlier chapters.

Consider first the matter of the computational precision. We restrict our attention to logarithms to the base 2, and we will assume that X and Y, the two factors in a multiplication, or the numerator and denominator in a division, have been put in a normalized form such that

$$1 \leq (|X|, |Y|) < 2$$

The number of shifts required to place the operands in this form constitutes the characteristic of the logarithm, and the mantissas M will lie in the range $0 \leq M < 1$. Over these limits, the logarithm function is as shown in Fig. 7.2. To generate our logarithms, we enter with an X value and read the $\log_2 X$ value (in a read-only memory, for example). To form a result we enter with a $\log_2 X$ value and read off an X value (the antilogarithm). Although the printed tables of logarithms employed in pencil-and-paper calculations can be used either to find the logarithm of a given number or to find the number corresponding to a given logarithm, the read-only memory (ROM) of the computer does not have this flexibility. We must use separate tables for these two functions. To find $\log_2 X$, we treat X as the address and enter a ROM of stored $\log_2 X$ values to find the desired logarithm. To find antilog Z (where Z is a logarithm), we use Z as an address to enter a ROM of stored antilogs. If n bits are needed in the antilog, then each quantum in the antilog range $1 \leq X < 2$ will be 2^{-n}. How many

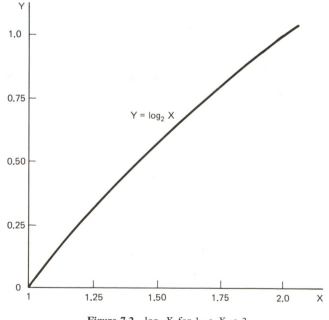

Figure 7.2 $\log_2 X$, for $1 \leq X \leq 2$.

bits must $\log_2 X$ contain so that two adjacent stored values of $\log_2 X$ will give this amount of discrimination in X? We may observe that the slope of the $\log_2 X$ versus X curve is smaller at $X = 2$ than at any point in the range 1 to 2. Thus the smallest increment in $\log_2 X$ that we must be able to represent is that slope times 2^{-n}. This gives

$$\text{smallest increment of } \log_2 X = \frac{1}{2 \ln 2} \times 2^{-n}$$

$$\simeq 1.44 \times 2^{-(n+1)}$$

Thus $(n + 1)$ bits in each log will be somewhat more precision than we need throughout the range. The ROM to store our antilog table will therefore have $2^{(n+1)}$ entries, each of n bits, for a total of $n \times 2^{(n+1)}$ bits. Each log table (we assume duplicate tables, so that $\log_2 X$ and $\log_2 Y$ may be simultaneously fetched) will have 2^k entries, each of which will have $(n + 1)$ bits, where k is the number of bits in each of X and Y. The total ROM capacity needed is therefore

$$2 \times (n + 1) \times 2^k + n \times 2^{(n+1)}$$

If, for example, $n = k = 8$, we require a capacity of

$$2 \times 9 \times 2^8 + 8 \times 2^9 = 8704 \text{ bits}$$

This is not an unreasonable number, in today's technology, but note that we must also provide the prenormalization circuitry and the adder to add the two logarithms (or to subtract them for division). Furthermore, all we have achieved is an 8-bit $\times$ 8-bit multiplication to form an 8-bit product. The necessary ROM size more than doubles for each increment of one bit in n and k, so that the memory size needed for multiplication of large numbers (40 to 60 bits) is quite out of the question for the technology of the foreseeable future.

Now consider the feasibility of generating the logarithms and antilogarithms when they are needed, instead of storing complete tables of both functions. We need a procedure that will be sufficiently accurate and that will be fast enough to compete with algorithmic multiplication and division procedures. An obvious candidate is to approximate the functions by straight-line segments, and the simplest (and crudest) such approximant is a single straight line, as shown in Fig. 7.3. It was pointed out by J. N. Mitchell, Jr. [7] that if X is normalized so that its most significant nonzero digit lies immediately left of the radix point (i.e., $1 \le X$ normalized < 2), then the fraction part of X constitutes $X - 1$, the single-straight-line approximant to $\log_2 X$. While this is a rather poor approximation for most purposes, it might conceivably be acceptable in some situations. It is a better approximation for X near 1 or 2 than in the center of the range. For example, if $X = 17$ and $Y = 31$, then $X \times Y = 527$. The binary forms of X_{norm} and Y_{norm} are 1.0001 and 1.1111, respectively. Thus

$$\log_2 X + \log_2 Y = 100.0001 + 100.1111$$

$$\log_2 (X \times Y) = 1001 = 9_{10}$$

$$X \times Y = 2^9 = 512$$

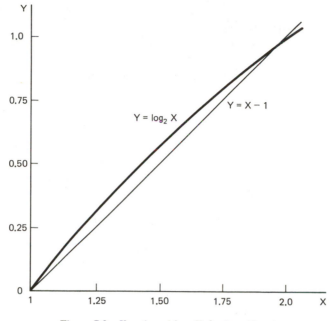

Figure 7.3 $X - 1$, and $\log_2 X$, for $1 \leq X \leq 2$.

using this crude approximation. If, however, we take $X = Y = 23$, we have $X \times Y = 529$. The binary forms are $X = Y = 10111$, which lead to

$$\log_2 X + \log_2 Y = 100.0111 + 100.0111$$

$$\log_2 (X \times Y) = 1000.111$$

$$X \times Y = 2^8 \times (1 + 7/8)$$

$$= 480$$

The error in this result is about 9.3%, while the previous example, involving numbers closer to the extremities of the range, had an error of about 2.8%.

The single-straight-line approximation is too crude for most computations, but the error can be reduced by using several straight lines. Even using only two straight lines would markedly improve the accuracy, but this would still not be enough for many purposes. Hall, Lynch, and Dwyer [9] have analyzed the use of one, two, four, and eight piecewise-linear approximants to the log curve, with the breakpoints chosen so as to minimize the mean-squared error, and have suggested that for certain real-time digital filtering applications these approximations may be sufficiently accurate.

Having adopted one or another of these ways of generating logarithms and anti-logarithms, we might reasonably consider executing our computations solely on logarithmic numbers. This makes addition and subtraction difficult operations, while facilitating multiplication and division. Several proposals [12–15] have been made for ways of circumventing this difficulty, at least over a limited range suited to a specific problem. In the *sign/logarithm* representation, due to Swartzlander and Alexopoulos [12,14], a number is represented by its sign and the logarithm of its scaled magnitude, where the scaling is chosen to assure that the logarithm will not be negative. Multiplication and division are executed, respectively, by addition or subtraction of the logarithm parts, with the result sign being set according to the signs of the operands. Since each operand is scaled by a factor k, multiplication of $A \times B$ involves $\log(kA) + \log(kB) = \log(k^2AB)$, and division produces $\log kA - \log kB = \log(A/B)$. Thus multiplication should be corrected by subtracting $\log k$ and division by adding $\log k$. To add two numbers A and B, in log form, we note that the sum S is

$$S = A + B$$
$$= A \times \left(1 + \frac{B}{A}\right)$$

The multiplication and division parts of this relation are easily carried out on numbers in log form, and a table of $[1 + (B/A)]$ can be provided to generate that part of the process. This technique has attractive possibilities for use in special-purpose processors where its accuracy is adequate and its cost not excessive for the purpose.

7.6 RESIDUE NUMBER SYSTEMS [16–28]

Fixed-base positional number systems, which are so familiar to us in everyday life and in most computer systems, have the severe disadvantage that arithmetic operations are slowed down by the necessity of carry propagation from each digit position to the next. Although (as we have seen) there are ways to speed up this carry propagation, it remains a problem requiring sequential actions rather than simultaneous ones. In the mid-1950s various workers in computer arithmetic devised arithmetic schemes based on representing a number by its residues with respect to a set of relatively prime moduli. Probably the first to develop such methods was Antonin Svoboda, who had at that time returned to his native Czechoslovakia after having spent the war years at the Radiation Laboratory at MIT. Governmental restrictions made it difficult for Svoboda's work to be published, but word-of-mouth information about it was conveyed to America by Howard Aiken and subsequently become widely known. Details were ultimately published (in the Czech language) in 1957.

Residue number systems can be easily explained by use of the congruence relationship. Two integers a and b are said to be "congruent with respect to the modulus m" (or equivalently, "congruent modulo m") if their difference is exactly divisible by m. The modulus m separates the integers into m residue classes, in each of which all the included integers are congruent to one another modulo m. For example, the modulus $m = 3$ forms the following three residue classes:

$$\{\cdots -6 \quad -3 \quad 0 \quad 3 \quad 6 \quad \cdots\}$$
$$\{\cdots -5 \quad -2 \quad 1 \quad 4 \quad 7 \quad \cdots\}$$
$$\{\cdots -4 \quad -1 \quad 2 \quad 5 \quad 8 \quad \cdots\}$$

Every integer falls into one or another of these three classes. The m smallest nonnegative elements (one from each residue class) are commonly known as the "residues" of their respective classes, and every integer is congruent modulo m to one of these residues. The congruence relation is indicated by the notation

$$a \equiv b \bmod m$$

We note that congruences with respect to the same modulus can be added. If we have

$$a \equiv b \bmod m$$

and

$$c \equiv d \bmod m$$

then

$$a - b = k_1 \times m$$

and

$$c - d = k_2 \times m$$

where k_1 and k_2 are integers. Thus

$$(a + c) - (b + d) = (k_1 + k_2) \times m$$

But this implies that the difference $(a + c) - (b + d)$ is exactly divisible by m, so we have

$$(a + c) \equiv (b + d) \bmod m$$

The sum modulo m of two residues X and Y is a number Z that is also one of the m residues. We may regard Z as being the least nonnegative element of the residue class into which $X + Y$ falls when we use ordinary arithmetic to add $X + Y$. For instance, $1 + 2 \equiv 0 \bmod 3$ since $1 + 2 = 3$ and 0 is the residue of the class in which 3 is found. Subtraction of congruences is also valid, and so is multiplication, which is merely repeated addition. Division, however, is undefined, since congruence is a relation among integers.

Consider now a set of k moduli $m_1, m_2, \cdots, m_k$ that are pairwise relatively prime. The integers $0, 1, 2, \cdots, (M - 1)$ (where $M = m_1 \times m_2 \times \cdots \times m_k$) have different residues with respect to these moduli and thus can be regarded as represented uniquely by those residues. For example, let $m_1 = 2$, $m_2 = 3$, and $m_3 = 5$. Then the integers 0 through 29 have the residues listed in Table 7.10, with respect to these three moduli.

The integer 30 has residues (0, 0, 0), just as does 0, and the cycle of values repeats for every 30 integers. This, of course, is exactly what we would expect from a finite number system, since infinitely many numbers will map onto each member of the finite set of representations, just as in the case of a fixed-base number system. The strikingly useful features of this residue encoding are:

1. It is an additive encoding: the representation of the sum of two integers is the sum of their representations.
2. The sum of the residues with respect to each modulus is quite independent of the residues in the other positions; that is, there is no phenomenon of carry from one position to the next.

These two features together make the residue number system an extremely attractive way to perform arithmetic rapidly in certain kinds of situations.

A number of difficulties must be noted, however. First, the magnitude comparison of two residue-coded numbers is difficult, because the set of residues conceals the magnitude from casual determination. For example, is "420" larger than "110"? The

TABLE 7.10

	Modulus		
Integer	$m_3 = 5$	$m_2 = 3$	$m_1 = 2$
0	0	0	0
1	1	1	1
2	2	2	0
3	3	0	1
4	4	1	0
5	0	2	1
6	1	0	0
7	2	1	1
8	3	2	0
9	4	0	1
10	0	1	0
11	1	2	1
12	2	0	0
13	3	1	1
14	4	2	0
15	0	0	1
16	1	1	0
17	2	2	1
18	3	0	0
19	4	1	1
20	0	2	0
21	1	0	1
22	2	1	0
23	3	2	1
24	4	0	0
25	0	1	1
26	1	2	0
27	2	0	1
28	3	1	0
29	4	2	1

TABLE 7.11

Integer	Modulus		
	5	3	2
0	0	0	0
1	1	1	1
2	2	2	0
3	3	0	1
4	4	1	0
5	0	2	1
6	1	0	0
7	2	1	1
8	3	2	0
9	4	0	1
10	0	1	0
11	1	2	1
12	2	0	0
13	3	1	1
14	4	2	0
−15	0	0	1
−14	1	1	0
−13	2	2	1
−12	3	0	0
−11	4	1	1
−10	0	2	0
−9	1	0	1
−8	2	1	0
−7	3	2	1
−6	4	0	0
−5	0	1	1
−4	1	2	0
−3	2	0	1
−2	3	1	0
−1	4	2	1

table shows 420 to be the residue code for 14_{10} and 110 to be the residue code for 16_{10}, so that 420 represents a smaller number than 110. In a fixed-base number system, we might make the comparison by subtracting one number from the other and ascertaining whether the result was positive or negative. In this case, $420 - 110 \rightarrow 310$, which our table shows to be the representation of $+28$. This strange outcome follows from the fact that we have not yet defined any negative numbers. Since the congruence relation defines residues with respect to both positive and negative integers, it is clear that we could let some of our $m_1 \times m_2 \times \cdots \times m_k$ encodings represent positive numbers and others represent negative numbers. It is reasonable to let the first half of our encodings represent positive numbers (and 0) and the others represent negative numbers, as tabulated in Table 7.11.

With these choices, we note that the negative of any integer in the range -15 to

+14 [or, more generally, $(-M/2)$ to $(M/2) - 1$] may be formed by "complementing" each residue with respect to its modulus. (The complement of x_j is defined to be $m_j - x_j$ for $x_j \neq 0$, and is 0 if $x_j = 0$.)

Now we are able to compare two numbers by subtracting them, to get a positive or a negative result, but we meet the second difficulty, which is that residue encoded numbers carry with them no obvious and simple indication of their signs. The sign of the number is not indicated by a single bit but is a function of all the residue digits. Of course, one way to handle the magnitude comparison problem and the sign determination problem is to decode the numbers to their conventional form, but this is self-defeating, and methods exist within the residue number system itself.

To *decode* a residue number, that is, to convert it to conventional form, we must solve a set of simultaneous congruences:

$$X \equiv a_1 \bmod m_1 \quad X \equiv a_2 \bmod m_2 \quad \cdots \quad X \equiv a_k \bmod m_k$$

where $m_1, m_2, \ldots, m_k$ are pairwise relatively prime. To do this, we make use of the Chinese remainder theorem, which enables us to determine a set of numbers y_j such that

$$X \equiv a_1 y_1 + a_2 y_2 + \cdots a_k y_k \bmod M$$

where $M = m_1 \times m_2 \times \cdots m_k$. The values y_j are defined to be $y_j = M_j \times M'_j$, where

$$M_j = \prod_{i=1, i \neq j}^{k} m_i$$

and

$$M_j \times M'_j \equiv 1 \bmod m_j$$

(i.e., M_j and M'_j are each other's multiplicative inverses). For example, given the set of moduli $m_1 = 7$, $m_2 = 5$, and $m_3 = 3$, we seek the number X whose residues are, respectively, 1, 3, and 1. We then find

$$M_1 = 5 \times 3 \qquad M_2 = 7 \times 3 \qquad M_3 = 7 \times 5$$

Then

$$15 \times M'_1 \equiv 1 \bmod 7 \qquad 21 \times M'_2 \equiv 1 \bmod 5 \qquad 35 \times M'_3 \equiv 1 \bmod 3$$

So

$$M'_1 = 1 \qquad M'_2 = 1 \qquad M'_3 = 2$$

and

$$y_1 = 15 \qquad y_2 = 21 \qquad y_3 = 70$$

From this we find that

$$X \equiv 15a_1 + 21a_2 + 70a_3 \bmod 105$$

$$\equiv 15 + 63 + 70$$

$$\equiv 148$$

The smallest nonnegative residue of X mod 105 is the value 43, which is readily seen to have the given residues 1, 3, and 1 with respect to the moduli 7, 5, and 3.

Obviously, the residue number system is not without drawbacks, since conversion to conventional form requires performing $\Sigma \, a_j \times y_j$. (The y_j need be calculated only once, of course.) Therefore, applications for residue arithmetic will be situations in which conversion to conventional form is needed only infrequently.

Conversion to conventional form is one way of determining the sign of a residue number. For instance, with moduli 7, 3, and 5, we could represent 105 different numbers, which might reasonably be those from -52 to $+52$. Any residue number that converts to a conventional number in the range $53 \leq X < 105$ is then identified as a negative number equal to $X - 105$. For instance, the number X whose residues mod 7, mod 5, and mod 3 are, respectively, 4, 3, and 1 converts by our procedure to the conventional number 88, which is then recognized as the negative number $X = -17$.

An obvious disadvantage of using conversion to conventional form as a means of sign detection is the need to provide facilities for doing arithmetic in conventional form, since the calculations cannot be executed by an arithmetic unit that performs only residue arithmetic. A method using only residue arithmetic converts the number to a mixed radix form in which the sign can be determined. Mixed radix forms express numbers as weighted sums:

$$X = \Sigma \, a_j \times W_j$$

but the ratios W_j / W_{j-1} are different for the various positions j. The mixed radix form we use here will express numbers as

$$X = \cdots + a_4 m_3 m_2 m_1 + a_3 m_2 m_1 + a_2 m_1 + a_1$$

For a set of N moduli $m_N, m_{N-1}, \ldots, m_1$ the leading term of X will be

$$a_N \prod_{i=1}^{N-1} m_i$$

Thus for moduli 7, 5, and 3, we could express numbers as

$$X = a_3 \times 15 + a_2 \times 3 + a_1$$

Clearly,

$$0 \leq a_1 \leq 2$$
$$0 \leq a_2 \leq 4$$
$$0 \leq a_3 \leq 6$$

Since sign detection requires a determination of which half of the number range contains our number X, it is helpful to let the last of our moduli be the value 2. If we then take

$$m_1 = 7$$
$$m_2 = 5$$
$$m_3 = 3$$
$$m_4 = 2$$

a residue number in terms of these four moduli can be converted to a mixed radix number

$$X = a_4 \times m_3 m_2 m_1 + a_3 \times m_2 m_1 + a_2 \times m_1 + a_1$$

$$X = a_4 \times 3 \times 5 \times 7 + a_3 \times 5 \times 7 + a_2 \times 7 + a_1$$

The digit a_4 will be 0 for numbers in the lower half of the range 0 through 209 and will be 1 for numbers in the upper half of that range, thus indicating the sign without explicitly forming the conventional representation.

To find the digits a_4, a_3, a_2, and a_1, we note that

$$X \equiv a_1 \bmod m_1$$

Hence a_1 is simply the first residue digit. We then find the residue form of $X - a_1$, with respect to the moduli m_4, m_3, and m_2. We now have

$$X - a_1 = a_4 \times m_3 m_2 m_1 + a_3 \times m_2 m_1 + a_2 \times m_1$$

Since m_1 is a divisor of $X - a_1$, we divide it out. This may be done by multiplying each residue of $X - a_1$ by the multiplicative inverse of m_1, after which we know the value of a_2. We repeat the process until we have found a_4.

Consider the following example:

	m_1	m_2	m_3	m_4	
	7	5	3	2	
Residues of $X =$	④	2	2	1	$a_1 = 4$
Subtract $a_1 = 4$:	-4	-4	-4	-4	
$X - a_1$	0	3	1	1	
Multiplicative inverse of 7 $=$		3	1	1	
Multiply these by $(X - a_1)$:		④	1	1	$a_2 = 4$
Subtract $a_2 = 4$		-4	-4	-4	
		0	0	0	
Multiplicative inverse of 5 $=$			2	1	
Multiply:			⓪	①	$a_3 = 0$
					$a_4 = 1$

Therefore: $X = 1 \times (3 \times 5 \times 7) + 0 \times (5 \times 7) + 4 \times (7) + 4$

$$= 137$$

Note that if we merely wish to find the sign, the final multiplication and addition to form 137 need not be carried out. The value $a_4 = 1$ is an indication that the number is negative (since it lies in the upper half of the range). Furthermore, all the operations needed to find a_4 can be done in residue arithmetic and do not require use of a conventional arithmetic unit. Because addition can be performed in the residue number system without the time delay required by carry propagation, it held promise of being a rapid way of executing arithmetic. As we have seen, however, sign detection poses difficulties, and conversion between residue and conventional form is also awkward. Further difficulties are encountered in overflow detection. The most difficult problem, however, is division, and there are no convenient ways of executing division in a simple and rapid manner. Thus residue arithmetic has found its principal application to be in situations where division is not required or where a complicated algorithm can be tolerated. A number of published papers carry the subject of residue arithmetic further than we will pursue it here.

7.7 p-ADIC NUMBERS [29–31]

The p-adic number representation is a generalized form of radix complement notation. A number is represented in p-adic form by an infinitely long sequence of digits, extending from a finite distance to the right of the radix point to infinitely far left, the digits being determined by a set of rules involving a base p, which is usually taken to be a prime number. Of course, the infinite length of such numbers makes them inappropriate for computation without some adjustment to make them finitely long, but several proposals have been made for truncating them in such a way as to take advantage of certain computational and representational features that they possess.

Before we state the general rules for forming p-adic representations, we will examine some specific cases using as a base $p = 10$. Although 10 is not a prime number, it will serve to introduce the ideas in a familiar form. All positive p-adic integers will be considered to be extended infinitely far left by 0's, as, for instance,

$$37 = \cdots 000 \cdots 037)_{\text{10-adic form}}$$

Whenever a digit pattern is repeated to the left, it is convenient to write that pattern just once, and to indicate the repeating pattern by a mark at its right. Thus we write

$$37 = 0'37$$

The negative of an integer is obtained by subtracting it from 0:

$$
\begin{array}{r}
\cdots 000000 \\
- \cdots 000037 \\
\hline
\cdots 999963 = 9'63
\end{array}
$$

In radix complement notation, using a fixed word length, -37 would be represented by some finite number of 9's to the left of the group 63, but in p-adic notation we have infinitely many $(p - 1)$'s at the left. Such numbers can be added and subtracted in the usual way. For instance:

	(a)		(b)		(c)	
	$13 = 0'13$		$24 = 0'24$		$24 = 0'24$
	$-76 = 9'24$		$-13 = 9'87$		$-24 = 9'76$
	$-63 = 9'37$		$11 = 0'11$		$0 = {}'0$

Multiplication of p-adic numbers, either positive or negative, is also carried out in the customary way. For example, we multiply $(-3) \times (-4)$ as follows:

$$-3 = \quad \quad 9'7$$
$$\times (-4) = \quad \times 9'6$$

$$\begin{array}{r} 9'8\,2 \\ 9'7\,3 \\ 9'7\,3 \\ \cdot \\ \cdot \\ \cdot \\ \hline \end{array}$$

$$\ldots 0\,0\,1\,2 = 0'12$$

Again:

$$-3 = \quad 9'7$$
$$\times (+4) = \quad 0'4$$
$$\overline{\quad \quad 9'88}$$

To find the multiplicative inverse of a number X, we seek a number X^{-1} such that $X \times X^{-1} = 1$. For instance, let us find the 10-adic inverse of $0'7$ by finding a number that multiplies $0'7$ to form $0'1$. The rightmost digit of X^{-1} will be 3, since $3 \times 7 = 21$. We subtract $0'21$ from $0'1$, and continue the process toward the left. The work is most conveniently arranged in the familiar style of longhand division, except that we work right to left, as follows:

$$
\begin{array}{r}
2\,8\,5\,7\,1\,4\,3 \\
0'7\,\overline{\big|\,\cdot\cdot\cdot 0\,0\,0\,0\,0\,0\,0\,1} \\
0'2\,1 \\
\hline
9'8\,0 \\
2\,8 \\
\hline
9'7\,0 \\
7 \\
\hline
9'0 \\
4\,9 \\
\hline
9'5\,0 \\
3\,5 \\
\hline
9'6\,0 \\
5\,6 \\
\hline
9'4\,0 \\
1\,4 \\
\hline
9'8\,0
\end{array}
$$

As in longhand division, we recognize at this point a repeating pattern, so that we may now write

$$\frac{1}{7} = 285714'3)_{\text{10-adic form}}$$

To verify our result, we may multiply it by 0'7:

$$
\begin{array}{r}
285714'3 \\
\times \qquad 0'7 \\
\hline
\cdots \qquad 000000\ 1 = 0'1
\end{array}
$$

Our reason for preferring a prime number for p is evident if we try to take the inverse of 4 by the same procedure, since there is no integer multiple of 4 that ends in 1, and similarly for 2, 5, 6, and 8. For the division process to be defined the rightmost digit of the divisor must be a nonzero number that is relatively prime to p. All rational numbers (i.e., numbers that can be expressed as the ratio of two integers) will have p-adic representations that have a repeating pattern of digits. Irrational numbers do not have such repeating groups, either in p-adic or in conventional form.

How can we proceed from the p-adic form, with its infinitely repeating pattern, to the conventional form? First we will denote our p-adic numbers by

$$d_{n+k}d_{n+k-1} \cdots d_{n+1}d_n d_{n-1} \cdots d_2 d_1 d_0$$

where the segment $d_n d_{n-1} \cdots d_2 d_1 d_0$ represents the nonrepeating part, and the k-digit group $d_{n+k} \cdots d_{n+1}$ represents the repeated pattern. The conventional form may now be formed by evaluating

$$N = \sum_{i=0}^{n} d_i p^i - \frac{1}{p^k - 1} \times \sum_{n+1}^{n+k} d_i p^i \tag{7.1}$$

For instance, we found that $1/7 = 285714'3$, for which we have $n = 0$ and $k = 6$. Then

$$N = 3 - \frac{1}{10^6 - 1} \times (2 \times 10^6 + 8 \times 10^5 + 5 \times 10^4 + 7 \times 10^3 + 1 \times 10^2 + 4 \times 10)$$

$$= 3 - \frac{2857140}{999999} = 0.\underline{142857}$$

The underbar is used to denote a pattern that repeats infinitely to the right. Note that the effect of dividing a k-digit group (285714) by $10^k - 1$ is to repeat that k-digit pattern infinitely to the right, and subtracting p times that group then from the nonrepeating part gives the number in conventional form.

As another example, consider

$$1/13 = 692307'7)_{\text{10-adic form}}$$

We again find $n = 0$ and $k = 6$, so that

$$N = 7 - \frac{10}{999999} \times (692307)$$

$$= 0.076923076\ldots = 0.\underline{076923}$$

p-adic systems using a prime number for p are more useful than the system based on 10. One way of forming a p-adic representation of the rational number a/b is to form b^{-1} using the base p, in the way we have already done, and then to multiply that value by a, again using the base p. For instance, let us find the 5-adic form of $3/8$ in this manner. First we form 8^{-1}:

$$
\begin{array}{r}
142 \\
8\overline{\big|\cdots 00001} \\
31 \\
\hline
4'20 \\
112 \\
\hline
4'330 \\
13 \\
\hline
4'20
\end{array}
$$

Pattern repeats, so $8^{-1} = 14'2)_{5\text{-adic}}$

Now we multiply by 3:

$$
\begin{array}{r}
14'2 \\
\times \quad 3 \\
\hline
\cdots 2\;21 \qquad \leftarrow \text{Carries} \\
\cdots 32321 \\
\hline
\cdots 03031 = 03'1, \text{ 5-adic form of } 3/8
\end{array}
$$

This result may be easily verified using (7.1).

Another procedure for obtaining the coefficients d_j of the p-adic expansion is due originally to Bachman [29] and is presented below in an algorithmic form. We let

$\alpha = a/b$, the rational number to be expanded

$\beta, \gamma = $ dummy variables of the algorithm

$p = $ prime number forming the base of the p-adic expansion

$c, d = $ integers not divisible by p

$n = $ an integer

The steps of the algorithm are as follows:

1. Initialize.
   (a) Set $\beta = \alpha$.
   (b) Find c, d, and n such that $\beta = (c/d) \times p^n$.
   (c) If $n > 0$, set $d_{n-1} \cdots d_2 d_1 d_0 = 0 \cdots 000$ and go to step 3.
2. Find c, d, and n such that $\beta = (c/d) \times p^n$.
3. Solve the congruence $d \times x_n \equiv 1 \bmod p$; then form $d_n = \text{residue } (c \times x_n) \bmod p$.
4. Form $\gamma = \beta - d_n \times p^n$. If $\gamma = 0$, set $d_i = 0$ for $i > n$; otherwise, set $\beta = \gamma$ and go to step 2.
5. Place p-adic point ("radix point") between d_{-1} and d_0; then EXIT.

As an example, we will use this procedure to find again the 5-adic form of $3/8$:

$$\text{I. 1. } \beta = 3/8 = 3/8 \times 5^0, \text{ where } c = 3$$
$$d = 8$$
$$n = 0$$
$$\text{3. } 8 \times x_0 \equiv 1 \bmod 5 \rightarrow x_0 = 2$$
$$d_0 = \text{res}(3 \times 2)_5 = 1$$
$$\text{4. } \gamma = 3/8 - 1 \times 5^0 = -5/8 = \beta$$
$$\text{II. 2. } \beta = -1/8 \times 5^1, \text{ where } c = -1$$
$$d = 8$$
$$n = 1$$
$$\text{3. } 8 \times x_1 \equiv 1 \bmod 5 \rightarrow x_1 = 2$$
$$d_1 = \text{res}(-1 \times 2)_5 = 3$$
$$\text{4. } \gamma = -5/8 - 3 \times 5^1 = -125/8 = \beta$$
$$\text{III. 2. } \beta = -125/8 = -1/8 \times 5^3, \text{ where } c = -1$$
$$d = 8$$
$$n = 3$$
$$\text{3. } 8 \times x_3 \equiv 1 \bmod 5 \rightarrow x_3 = 2$$
$$d_3 = \text{res}(-1 \times 2)_5 = 3$$

Continuing the process generates $d_{\text{odd}} = 3$, with the intervening values $d_{\text{even}} = 0$. Thus we find, as before, that

$$\frac{3}{8} = 03\overline{1})_{5\text{-adic}}$$

We may now broaden the scope of our p-adic representations by noting that the "p-adic point" (or radix point) need not be at the right of our digit group. As the preceding algorithm indicates, it is possible for n to have negative values at the start of the procedure, if the denominator d contains one or more multiples of the base. For example, if we seek the 5-adic expansion of $\frac{7}{15}$, we recognize that this is $\frac{7}{3} \times \frac{1}{5}$. We may either form the 5-adic expansion of $\frac{7}{3}$ and then shift it one position to the right, to multiply by $\frac{1}{5}$, or we may alternatively start our recursive algorithm by setting $c = 7$, $d = 3$, and $n = -1$, so that $\alpha = \frac{7}{3} \times 5^{-1}$. In either case, we would find that

$$\frac{7}{15} = 13'.4)_{\text{5-adic}}$$

In general, then, p-adic numbers may be multiplied by positive or negative powers of the base p by being shifted left or right just as we do with conventional numbers. Thus:

$$7/15 \times 5 = 13'4)_{\text{5-adic}}$$

$$7/15 \times 5 \times 5 = 13'40)_{\text{5-adic}}$$

$$7/15 \times 5^{-1} = 13'13.4 \times 5^{-1} = 13'1.34 = 31'.34$$

$$7/15 \times 5^{-3} = 31'.3134$$

It has been noted by Horspool and Hehner [31] that multiplication by powers of the base can be indicated by attaching an exponent to the significant digits of the expansion, just as we do in floating-point notation. Thus we could write:

$$7/15 = 13'4 \ E(-1)$$

$$7/15 \times 5 = 13'4 \ E0$$

$$7/15 \times 5 \times 5 = 13'4 \ E1$$

$$7/15 \times 5^{-1} = 13'4 \ E(-2)$$

$$7/15 \times 5^{-3} = 13'4 \ E(-4)$$

$$= 13'.4 \ E(-3)$$

$$= 1.3'4 \ E(-4)$$

$$= .13'134 \ E1$$

$$= 13'40 \ E(-5)$$

As the last example shows, there are many equivalent forms for a given number, just as in floating-point notation. The "normalized" form can be regarded as the one having the shortest significand and with the radix point at the right. The normalized 5-adic form for $\frac{7}{15} \times 5^{-3}$ is thus $13'4 \ E(-4)$.

7.8 SIGNED-DIGIT NUMBERS: AN EXTENSION OF THE FIXED-BASE SYSTEM [32–34]

Some interesting and useful variations of the fixed-base number system are possible if we use digit sets other than the familiar set

$$d = \{0, 1, 2, \ldots, (r - 1)\}$$

It is even somewhat surprising to note that for any radix r there exists an infinite number of digit sets. For instance, the following unusual set is possible for radix $r = 10$:

$$D = \{0, 1, 20, 21, 40, 41, 60, 61, 80, 81\}$$

Using this set, we would represent the integers 0 through 10 as follows:

0	$\longleftrightarrow$	0
1	$\longleftrightarrow$	1
2	$\longleftrightarrow$	$0.(20) = 0 + 20 \times 10^{-1}$
3	$\longleftrightarrow$	$0.(20) = 1 + 20 \times 10^{-1}$
4	$\longleftrightarrow$	$0.(40)$
5	$\longleftrightarrow$	$1.(40)$
6	$\longleftrightarrow$	$0.(60)$
7	$\longleftrightarrow$	$1.(60)$
8	$\longleftrightarrow$	$0.(80)$
9	$\longleftrightarrow$	$1.(80)$
10	$\longleftrightarrow$	10

and it may be entertaining to the reader to devise some other sets that are possible.

As a more practical matter, however, sets that include negative as well as positive digits have special interest to us. If the radix r is an odd number, we define a balanced digit set as the r-member set

$$B = \left\{ -\frac{r-1}{2}, -\frac{r-3}{2}, \ldots, -1, 0, 1, \ldots, +\frac{r-1}{2} \right\}$$

For instance, if $r = 3$ (the ternary system), the digit set is $B_3 = \{-1, 0, 1\}$. [Note that if the number of elements in the digit set equals the radix r, the set is nonredundant. We used this same set for recoding binary multipliers (in Chapter 4), but there the representation was a redundant one, since a binary number can be expressed in more than one way with that digit set.]

It is an easy matter to convert a positive number expressed in positive digits $\{0, 1, \ldots, (r - 1)\}$ to a form expressed in the balanced digit set. Let the number X be

$$X = d_n d_{n-1} \cdots d_1 d_0$$

Then:

1. Initialize
$$j \leftarrow 0$$
$$c_j \leftarrow 0$$

2. If $(d_j + c_j)$ is not a member of the balanced set, then
$$d_j \leftarrow -(r - d_j - c_j)$$
$$c_{j+1} \leftarrow 1$$

3. If $j = n$, then
$$d_{j+1} \leftarrow c_{j+1}$$
$$\rightarrow \text{EXIT}$$

    else
$$j \leftarrow j + 1$$
$$\rightarrow 2$$

For example, the base 3 number 122 is transformed by this procedure to $1\bar{1}0\bar{1}$. Conversion of a number from the balanced digit set form, or more generally from a signed-digit form (whether balanced or not) can be done by subtracting the weighted sum of the negative digits from the weighted sum of the positive digits. Thus

$$1\bar{1}0\bar{1} = 1000 - 0101$$
$$= 122$$

where of course the subtraction is done in base 3.

    The sign of a number expressed in terms of a signed-digit set is the same as the sign of the most significant nonzero digit. However, negating a number requires an operation on every digit, since there is no explicit sign separate from the magnitude. If the digit set is balanced, this operation is a simple sign reversal for every nonzero digit. If the digit set is not a balanced one, sign reversal may produce a digit not in the allowed set. Such a digit must then be recoded into the allowed set. For instance, an unbalanced set for radix $r = 10$ is

$$\{-6, -5, -4, -3, -2, -1, 0, 1, 2, 3\}$$

With that set, the number $X = 293728$ is written as $30\bar{6}33\bar{2}$. Negating that number digit by digit produces a positive 6, which must then be recoded as $1\bar{4}$. The final result would be $31\bar{4}33\bar{2}$. To avoid this inconvenient step in the negating process, signed-digit sets are usually selected so as to be balanced.

    A potential advantage of a balanced digit set is in multiplication, since only $(r - 1)/2$ multiples of the multiplicand need be furnished.

7.8.1 Redundant Signed-Digit Numbers

The introduction of redundancy into a signed-digit set can be useful. We have already seen the use of $\{-1, 0, 1\}$ in the Booth multiplication procedure. Similarly, pseudo-quotients for $r > 2$ were expressed in a signed-digit form using the set $\{-(r - 1), -(r - 2), \ldots, -1, +1, \ldots, (r - 1)\}$, a set having $(2r - 1)$ symbols. Another interesting form of redundant signed digits is the one devised by Avizienis [32,33] with the objective of defining an addition process free from carry propagation. Our discussion will follow his treatment, with some changes in notation to assure consistency with earlier chapters.

As with conventional fixed-base systems, a signed-digit number having $(n + 1)$ integer digits and m fraction digits has the algebraic value

$$Z = \sum_{i=-m}^{+n} z_i r^i$$

where the radix r is a positive integer and each digit z_i is chosen from the allowed set of positive and negative digit values. We require also that the algebraic value $Z = 0$ have a unique representation and that transformations exist between our signed-digit form and conventional form over the entire range of possible values of Z.

Carry propagation is absent if the number system allows what may be called "totally parallel" addition; that is, each sum digit is a function only of summand digits in that position and in the adjacent position. Figure 7.4 shows the relations required by this condition. The digits of X and Y are inputs to the A boxes, and the sum digits are obtained from the B boxes. Each sum digit S_i is thus a function only of x_i, y_i, x_{i-1}, and y_{i-1}. The relations are

$$x_i + y_i = r \times t_{i+1} + w_i \qquad \text{A boxes}$$

$$S_i = w_i + t_i \qquad \text{B boxes}$$

where the "+" designates arithmetic addition. x_i, y_i, and S_i may assume any of the allowed values for our signed-digit set, and t_i (the "transfer" digit) may be any one of $\{-1, 0, +1\}$.

In order that there may be a unique representation for the algebraic value zero, we must have $|z_i| \leq r - 1$. Otherwise, if we could have a digit $z_i = r$, we could represent zero both by $1r$ and by 00. Then, since $|S_i| \leq r - 1$ and t_i could be ± 1, we must have $|w_i| \leq r - 2$. This limits r to values of 3 or greater, since $r = 2$ gives 0 as the only possible value of w_i. (However, it should be noted that radix 2 numbers may

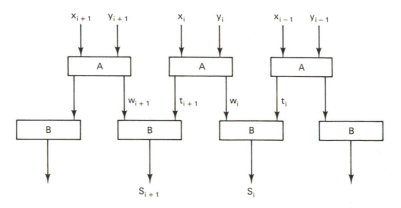

Figure 7.4   Totally parallel addition.

be expressed in signed-digit form and still have the property of totally parallel addition if they are expressed in canonical form, that is, the form in which nonzero digits are separated by at least one zero digit. Unfortunately, the sum of two canonical numbers is not necessarily a canonical number.)

To transform numbers between signed-digit and conventional forms, w_i must be capable of at least r successive values from the set $\{-(r-2), -(r-3), \ldots, -1, 0, 1, \ldots, (r-2)\}$. $[w_i$ cannot be either $(r-1)$ or $-(r-1)$, because these are the limiting values for $S_i = w_i + t_i.]$ For convenience, we choose a set that is symmetric about 0. For an odd radix $r_o \geq 3$, we take $w_{max} = -w_{min} = \frac{1}{2}(r_o - 1)$, giving as the minimum set of digit representations the $(r_o + 2)$ integers

$$\{-1/2(r_o + 1), \ldots, -1, 0, 1, \ldots, 1/2(r_o + 1)\}$$

For an even radix $r_e \geq 4$, we need $r_e + 3$ integers in the set, to assure that both extreme elements have additive inverses. We use as the minimum set

$$\{-(1/2\ r_e + 1), \ldots, -1, 0, 1, \ldots, +(1/2\ r_e + 1)\}$$

Thus we find the following digit sets possible:

Radix 3: $\{-2, -1, 0, 1, 2\}$ no others

Radix 4: $\{-3, -2, -1, 0, 1, 2, 3\}$ no others

Radix 5: $\{-3, -2, -1, 0, 1, 2, 3\}$

             $\{-4, -3, -2, -1, 0, 1, 2, 3, 4\}$

Radix 6: $\{-4, -3, -2, -1, 0, 1, 2, 3, 4\}$

             $\{-5, -4, -3, -2, -1, 0, 1, 2, 3, 4, 5\}$

Radix 10: $\{-6, -5, -4, -3, -2, -1, 0, 1, 2, 3, 4, 5, 6\}$ 13 elements

            $\{-7, -6, -5, -4, -3, -2, -1, 0, 1, 2, 3, 4, 5, 6, 7\}$ 15 elements

            $\{-8, -7, -6, -5, -4, -3, -2, -1, 0, 1, 2, 3, 4, 5, 6, 7, 8\}$
17 elements

            $\{-9, -8, -7, -6, -5, -4, -3, -2, -1, 0, 1, 2, 3, 4, 5, 6, 7, 8, 9\}$
19 elements

The rules given earlier for conversion between conventional and signed-digit forms are valid also for these redundant digit sets. However, since the sets are redundant, there are several forms possible. For instance, the number 656, with radix $r = 10$, can be represented in the 13-element set as 656, or 66$\bar{4}$, or 1$\bar{4}$56, or 1$\bar{4}$6$\bar{4}$, or 1$\bar{3}$44. It is reasonable to select that form which would appear at the output of our totally parallel adder structure of Fig. 7.4 if the conventional form of the number is added to 0. The outputs are equivalent to those obtained from the following algorithm, where again we initially take X as positive. For each i in turn:

1. If $x_i \geq$ largest number of the redundant digit set, then $w_i \leftarrow -(r - x_i)$ and $t_{i+1} \leftarrow 1$, else $w_i \leftarrow x_i$ and $t_{i+1} \leftarrow 0$.
2. $S_i \leftarrow w_i + t_i$.

As before, if X is a negative number, we then change the signs of all digits S_i. Applying this procedure to our example 656 would yield $1\overline{4}6\overline{4}$.

Although the redundant signed-digit system has the potential for high-speed operation, it has the disadvantage of requiring more complex circuitry and more storage space than needed for conventional numbers, and designers have not found the advantages to be worth the costs. Nevertheless, it may well prove useful where these costs can be justified.

7.9 REPRESENTATION OF COMPLEX NUMBERS [35–38]

The usual practice in manipulating complex numbers is to maintain separate representations for the real part and for the imaginary part, with the various arithmetic procedures then being programmed in the conventional way by appropriate software. Various proposals have been made, however, for a more concise representation in which a single group of digits forms a representation of both the real and the imaginary parts, allowing the programmer to be relieved of the problem of dealing with them separately. These proposals typically involve choosing as a "base" a complex or an imaginary number and then expressing the complex quantity as a weighted sum of powers of that base.

Consider the use of $2i$ as a base (where we use i to indicate $\sqrt{-1}$. Then the weights that we associate with position j are $(2i)^j$, or

j:	$\cdots$	5	4	3	2	1	0	-1	-2	-3	-4	$\cdots$
Weights:		$32i$	$+16$	$-8i$	-4	$2i$	1	$-\frac{1}{2}i$	$-\frac{1}{4}$	$\frac{1}{8}i$	$\frac{1}{16}$	

The weights associated with even-numbered positions are:

j:	$\cdots$	6	4	2	0	-2	-4	$\cdots$
Weights:		-32	$+16$	-4	1	$-\frac{1}{4}$	$+\frac{1}{16}$	

and those associated with odd-numbered positions are:

j:	5	3	1	-1	-3	$\cdots$
Weights:	$+32i$	$-8i$	$+2i$	$-\frac{1}{2}i$	$+\frac{1}{8}i$	$\cdots$
	16	-4	1	$-\frac{1}{4}$	$+\frac{1}{16}$	$\cdots) \times 2i$

The even-numbered positions all have real weights, which are related to each other by the rules for base $= -4$. Similarly, the odd-numbered positions all have imaginary weights, which after factoring out the common $2i$, are also related as in base $= -4$. Thus, to express any complex number in this format, we write its real part as an expansion in terms of base (-4) and we write half its imaginary part as an expansion in base (-4). The two parts may then be interleaved to give a single digit group. For instance, the complex number $9 + 7i$ has for its real part $121_{(-4)}$, that is,

$$9_{10} = 1 \times (-4)^2 + 2 \times (-4)^1 + 1 \times (-4)^0$$
$$= 1 \times (2i)^4 + 2 \times (2i)^2 + 1 \times (2i)^0$$

$7i$ can be written as

$$7i = 3.5 \times 2i = 130.2_{(-4)} \times 2i$$

since

$$7i = (16 - 12 - 1/2) \times 2i$$
$$= [(-4)^2 + 3 \times (-4)^1 + 2 \times (-4)^{-1}] \times 2i$$
$$= [(2i)^4 + 3 \times (2i)^2 + 2 \times (2i)^{-2}] \times 2i$$
$$= (2i)^5 + 3 \times (2i)^3 + 0 \times (2i)^1 + 2 \times (2i)^{-1}$$

Thus

$$(9 + 7i)_{10} = 113201.2$$

Addition of numbers in this form involves possible carries from even-numbered position to even-numbered position and from odd to odd. Thus, for addition at least, one might as well as handle the real and the imaginary parts separately. Because the effective base is negative, the carries will be opposite in sign to the position in which they arrive. (Thus a carry from position j should cause d_{j+2} to decrease by 1. If d_{j+2} was 0 before the arrival of the carry, then d_{j+4} should be increased by 1. This is a simple generalization of the carry concept for negabinary arithmetic.)

Here is an addition example:

$$5_{10} = 1\ 0\ 3\ 0\ 1_{(2i)}$$

$$11_{10} = 1\ 0\ 2\ 0\ 3_{(2i)}$$

$$\overline{1}\quad \overline{1} \quad \longleftarrow \text{Carries}$$

$$16\ \ = 1\ 0\ 0\ 0\ 0_{(2i)}$$

Multiplication can be handled either by treating the parts separately, or in a unified way in which we pay careful attention to carries. For instance,

$$5 + i \quad = 1\ 0\ 3\ 1\ 1.2_{(2i)}$$
$$\times\ 6 + 2i = 1\ 0\ 3\ 1\ 2.0_{(2i)}$$

$$28 + 16i$$

$$\bar{1} \qquad \bar{1}$$

$$2\ 0\ 2\ 2\ 2\ 0\ 0 \qquad \text{First partial product}$$

$$1\ 0\ 3\ 1\ 1\ 2 \qquad\qquad \text{Second partial product}$$

$$\bar{2} \qquad \bar{1}$$

$$3\ 0\ 1\ 3\ 3\ 2 \qquad\qquad \text{Third partial product}$$

$$\underline{1\ 0\ 3\ 1\ 1\ 2\ 0} \qquad\qquad \text{Last partial product}$$

$$0\ 0\ 0\ 1\ 2\ 2\ 1\ 0\ 0.0\ 0 \qquad \text{Result}$$

The result may be recognized as the encoding of $28 + 16i$ in the base $(2i)$ format.

An advantage of this encoding is its compactness when expressed in binary form, since each base (-4) digit requires exactly two binary digits for its representation. We could achieve a similar compactness by using $4i$ as a base, in which case each digit of our representation would be a hexadecimal digit, representable by four binary digits. Thus our positional weights would be:

$j = \cdots$	5	4	3	2	1	0	-1	$-2\ \cdots$
Weights =	$(4i)^5$	$(4i)^4$	$(4i)^3$	$(4i)^2$	$(4i)^1$	$(4i)^0$	$(4i)^{-1}$	$(4i)^2$
	$1024i$	256	$-64i$	-16	$4i$	1	$-1/4i$	$-1/16$
Real weights =		256		-16		$+1$		$-1/16$
Imaginary weights =	$1024i$		$-64i$		$+4i$		$-1/4i \cdots$	
=	256		-16		$+1$		$-1/16 \cdots) \times 4i$	

To achieve octal encodings, we could use $2\sqrt{2i}$ as the base. This would enable us to encode the real part and the imaginary part in terms of base 8.

Another choice for the base is the complex number $(i - 1)$ or the number $(-i - 1)$. Note that $(i - 1)^4 = -4$, $(i - 1)^8 = +16$, and so on through powers of (-4). Thus we can express real integers in terms of this negative base, using digits from the set $\{0, 1, 2, 3\}$ as coefficients of $(i - 1)^{4k}$. We note also that 0, 1, 2, and 3 may be encoded in terms of powers of $(i - 1)$ as follows:

Integer	$(i-1)^3 =$ $2 + 2i$	$(i-1)^2 =$ $-2i$	$(i-1)^1$	$(i-1)^0 =$ 1
0	0	0	0	0
1	0	0	0	1
2	1	1	0	0
3	1	1	0	1

Thus each of our base (-4) digits may be encoded as a 4-bit group.

A similar situation holds for the imaginary numbers. Since $(i - 1)^6 = 8i$, and $(i - 1)^{10} = -32i$, and so on, we see that those terms are of the form $8i \times (-4)^k$. That is,

$$\cdots \;(i-1)^{18}, \;\;(i-1)^{14}, \;\;(i-1)^{10}, \;\;(i-1)^6, \;\;(i-1)^2, \;\;(i-1)^{-2}, \;\;(i-1)^{-6} \;\cdots$$

are equal, respectively, to

$$(8i)(-4)^3, \;\; 8i(-4)^2, \;\; 8i \times -4, \;\; 8i, \;\; 8i \times (-4)^{-1}, \;\; 8i \times (-4)^{-2}, \;\; 8i \times (-4)^3 \;\cdots$$

Taking out the common factor $8i$, the weights are then found to be in the ratio

$$(-4)^3 \;\; (-4)^2 \;\; (-4)^1 \;\; (-4)^0 \;\; (-4)^{-1} \;\; (-4)^{-2} \;\; (-4)^{-3} \;\cdots] \times 8i$$

Thus $\frac{1}{8}$ of the imaginary part of a number may be encoded in base (-4). For instance, $36i$ may be encoded by dividing by $8i$ to get 4.5_{10}, which, in base (-4), is 131.2. We multiply the $8i$ factor in again, to get:

$$\text{Weights:} \quad (i-1)^{14} \quad (i-1)^{10} \quad (i-1)^6 \quad (i-1)^2$$

$$\text{Digits:} \quad 1 \qquad\quad 3 \qquad\quad 1 \qquad\quad 2$$

We may use the same 4-bit encodings as before, with the 4-bit group $A_{j+3} A_{j+2} A_{j+1} A_j$ replacing the base (-4) digit in position j, and with zeros in position 1 and 0. Thus our group 1312 may be written as:

$$\text{Position } j = 17 \; 16 \; 15 \; 14 \mid 13 \; 12 \; 11 \; 10 \mid 9 \; 8 \; 7 \; 6 \mid 5 \; 4 \; 3 \; 2 \mid 1 \; 0$$

$$\text{Base } (-4) \text{ digit:} \qquad\qquad 1 \mid \qquad\quad 3 \mid \qquad 1 \mid \qquad\quad 2 \mid$$

$$\text{Binary form:} \quad 0 \; 0 \; 0 \; 1 \mid 1 \; 1 \; 0 \; 1 \mid 0 \; 0 \; 0 \; 1 \mid 1 \; 1 \; 0 \; 0 \mid 0 \; 0$$

To execute arithmetic in this number system, we need rules for addition. These are simply:

$$0 + 0 = 0$$

$$1 + 0 = 0 + 1 = 1$$

$$1 + 1 = 1\,1\,0\,0$$

That is, adding two 1's in column j produces a 0 in that column and carries of 1 into columns $(j + 2)$ and $(j + 3)$.

Now let us carry out a simple multiplication in this number system. We will perform $(i + 2) \times i$, where the factors are represented as:

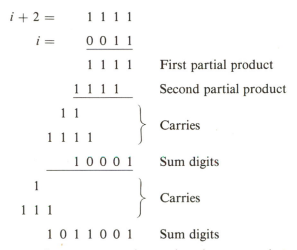

$$i + 2 = \quad 1\ 1\ 1\ 1$$
$$i = \quad\ \ 0\ 0\ 1\ 1$$

1 1 1 1	First partial product
1 1 1 1	Second partial product
1 1	
1 1 1 1	Carries
1 0 0 0 1	Sum digits
1	
1 1 1	Carries
1 0 1 1 0 0 1	Sum digits

This is as far as we need to propagate the carries, since successive assimilations will simply form more 0's at the left. This occurs because the pattern at the left

$$1\ 1\ 1$$
$$+ \quad 1\ 1$$

is identically zero.

The reader is urged to confirm his understanding of this number system by repeating the example of $(5 + i) \times (6 + 2i)$, after first expressing the two factors in base $(i - 1)$ format.

The two number systems that we have examined were proposed by D. E. Knuth [35] and by W. Penney [36]. Although they are interesting examples of unusual representations, they have not proved to be of significant practical value. Another technique, suggested by W. N. Holmes [37], uses a base 3 encoding of each digit position of a complex number and would seem to offer simpler hardware implementation than the complex base. In this method, each digit position encodes the real part and the imaginary part in a ternary notation that represents 0, $+1$, or -1 as the size of the real part and 0, $+1$, or -1 as the size of the imaginary part. For instance, we might use

Digit	3^1	3^0	Real part	Imaginary part
0	0	0	0	0
1	0	1	1	0
2	0	2	-1	0
3	1	0	0	1
4	1	1	1	1
5	1	2	-1	1
6	2	0	0	-1
7	2	1	1	-1
8	2	1	-1	-1

Each pair of these ternary digits (or each single digit $\{0, 1, \ldots, 8\}$) represents a single coefficient d_j in the base 3 expansion of a complex number:

$$\cdots d_3 \times 3^3 + d_2 \times 3^2 + d_1 \times 3^1 + d_0 \times 3^0 + d_{-1} \times 3^{-1} + \cdots$$

For instance,

$$\underset{d_2}{\underbrace{12}} \quad \underset{d_2}{\underbrace{01}} \quad \underset{d_0}{\underbrace{11.}} \quad \underset{d_{-1}}{\underbrace{02}}$$

in this format is the representation of

$$(-1 + i) \times 3^2 + (1 + i0) \times 3^1 + (1 + i1) \times 3^0 + (-1 + i0) \times 3^{-1}$$

$$= \overline{1}11.\overline{1} + i(101.0)$$

$$= -5\tfrac{1}{3} + i10$$

The rules for addition in this format are best represented by an addition table. Since our assignments of ternary pairs to the nine possible combinations of values of the real and imaginary components were quite arbitrary, we would not expect any logically simple rules to exist for forming sum and carry digits. Table 7.12, expressed in terms of the digit set $\{0, 1, \ldots, 8\}$ shows the results of forming $X + Y$ and $X - Y$.

To find $X + Y$, find the entry at column X and row Y as listed on the left. To find $X - Y$, again enter at column X, but use row $-Y$ as listed on the right. For instance, $3 + 4$ corresponds to $(0 + i) + (1 + i)$, which is $(1 + 2i)$. The $2i$ component is equal to $(1\overline{1})_3 \times i$, so $(1 + 2i)$ is $(0 + i) \times 3^1 + (1 - i) \times 3^0$, which we represent as 37.

The multiplication table is shown in Table 7.13.

TABLE 7.12 ADDITION AND SUBTRACTION

Y	\multicolumn{8}{c}{X}	−Y
	1	2	3	4	5	6	7	8	
1	12	0	4	15	3	7	18	6	−2
2	0	21	5	3	24	8	6	27	−1
3	4	5	36	37	38	0	1	2	−6
4	15	3	37	48	36	1	12	0	−8
5	3	24	38	36	57	2	0	21	−7
6	7	8	0	1	2	63	64	65	−3
7	18	6	1	12	0	64	75	63	−5
8	6	27	2	0	21	65	63	84	−4

TABLE 7.13 MULTIPLICATION

1	2	3	4	5	6	7	8
2	1	6	8	7	3	5	4
3	6	2	5	8	1	4	7
4	8	5	36	21	7	12	63
5	7	8	21	63	4	36	12
6	3	1	7	4	2	8	5
7	5	4	12	36	8	63	21
8	4	7	63	12	5	21	36

7.10 *SOME UNUSUAL AND UNUSEFUL NUMBER SYSTEMS*

The ingenuity of man knows no limits, and the number of ingenious schemes that can be proposed for representing numbers seems also quite without limit. Some of these methods have a degree of intellectual charm, but merely to be able to represent numbers in some clever and unusual fashion does not assure that conversion to and from that form will be simple, or that the representation will possess useful manipulative advantages in executing arithmetic. As examples, we consider the binomial number system [39] and the Fibonacci number system [40].

For the *binomial number system,* we assume a positional notation of n digits $d_n d_{n-1} d_{n-2} \cdots d_1$, and we require that the digits be related by

$$d_n > d_{n-1} > d_{n-2} > \cdots > d_2 > d_1 \geq 0$$

Any positive integer N within the range of the system is represented by an n-digit group whose digits are chosen with this constraint and in such a way that

$$N = \binom{d_n}{n} + \binom{d_{n-1}}{n-1} + \cdots + \binom{d_2}{2} + \binom{d_1}{1}$$

[The notation $\binom{a}{b}$ is the usual symbol for the binomial coefficients, and we compute its value by $a!/(a-b)!b!$.] For instance, in a three-digit system the first 20 integers would be represented as in Table 7.14. Note that this table shows:

$$1 \text{ group having } d_3 = 2$$

$$3 \text{ groups having } d_3 = 3$$

$$6 \text{ groups having } d_3 = 4$$

$$10 \text{ groups having } d_3 = 5$$

This progression is itself the binomial sequence $\binom{2}{2}$, $\binom{3}{2}$, $\binom{4}{2}$, $\binom{5}{2}$, and if we continue the table to $d_3 = k$, we would find $\binom{k}{2}$ groups starting with that value k.

How many different numbers can we encode in this way, with n digits in our

TABLE 7.14

Integer	d_3	d_2	d_1
0	2	1	0
1	3	1	0
2	3	2	0
3	3	2	1
4	4	1	0
5	4	2	0
6	4	2	1
7	4	3	0
8	4	3	1
9	4	3	2
10	5	1	0
11	5	2	0
12	5	2	1
13	5	3	0
14	5	3	1
15	5	3	2
16	5	4	0
17	5	4	1
18	5	4	2
19	5	4	3

representation and with some maximum value k for d_n? For $n = 3$ and $k = 5$, this number is the sum of the series $\binom{2}{2} + \binom{3}{2} + \binom{4}{2} + \binom{5}{2}$, and for general k, the number of three-digit representations is

$$\sum_{2}^{k} \binom{j}{2} = \binom{k+1}{3}$$

If, for example, we represent each of the three digits by a 4-bit binary code, we could let d_3 take on the 16 values 2 through 17, d_2 the 16 values 1 through 16, and d_0 the 16 values 0 through 15. Thus $k = 17$, and the number of representations available is $\binom{18}{3}$ = 816. Of course, this is not an efficient representation, since our 12 bits are capable of 4096 combinations.

With larger n, we can represent more numbers, and it can be shown without much difficulty that the number of n-digit representations possible with a maximum value of k for d_n is given by $\binom{k+1}{n+1}$.

Having shown that it is indeed possible to represent numbers in this fashion, we have gone about as far as we can easily go. Conversion of numbers into this form is a nasty algebraic process. For instance, to find $d_4 d_3 d_2 d_1$, for $N = 1800$, we solve first for d_4, which is the largest integer x such that $\binom{x}{4} \leq 1800$:

$$\frac{x!}{(x-4)!4!} \leq 1800$$

$$\frac{1}{4!} x(x-1)(x-2)(x-3) \leq 1800$$

As a reasonable starting point, we choose $x = \lceil \sqrt[4]{1800 \times 4!} \rceil$ since we know that x must be an integer. We find as a trial value $x = 15$. By inspection, we find that $x = 15$ satisfies $\binom{x}{4} \leq 1800$, but $x = 16$ does not. Subtracting out $\binom{15}{4}$, we can then evaluate d_3 in a similar way. These procedures do not lend themselves to simple implementation in the computer, but problems even more formidable occur when we start to perform arithmetic with these representations. The principal difficulty is that now that we have mapped the natural numbers into this new domain, we do not have available any simple relation among the elements of the new domain to facilitate the arithmetic operations. Thus we are forced to conclude that the binomial number system serves no useful computational purposes, and its chief claim on our interest is as a didactic exercise in binomial coefficients.

The *Fibonacci number system* (so-called by Knuth) is another remarkable and remarkably useless number system. Recall that the Fibonacci numbers F_j are defined by the recursive relation

$$F_j = F_{j-1} + F_{j-2}$$

with the initial values $F_0 = 0$ and $F_1 = 1$. Thus the series of Fibonacci numbers starts out as follows:

$$j = 0 \ 1 \ 2 \ 3 \ 4 \ 5 \ 6 \ \ 7 \ \ 8 \ \ 9 \ 10 \quad \cdots$$

$$F_j = 0 \ 1 \ 1 \ 2 \ 3 \ 5 \ 8 \ 13 \ 21 \ 34 \ 55 \quad \cdots$$

One can spend many pleasant hours playing with this series and its various forms, and much has been written about it [41,42]. For our present purposes, however, it is enough to note that the positive integers can be encoded by a set of digits $d_n d_{n-1} \cdots d_1$, where each digit is a Fibonacci index j and the value of the number represented is given by

$$N = F_{d_n} + F_{d_{n-1}} + F_{d_{n-2}} + \cdots + F_{d_1}$$

If we restrict the digits d_i so that for all i from 1 to n, $d_i \geq d_{i-1} + 2$, then the encoding is unique. (However, we will allow the leftmost digits of our n-digit group to be 0.) The first several positive integers may be thus encoded as in Table 7.15.

To perform the encoding of N, we first find F_k, the largest Fibonacci number not greater than N. If $N = F_k$, our encoding consists of the single digit k, the index of F_k. If not, we form $N_1 = N - F_k$, which is certainly smaller than F_{k-1}, but not necessarily smaller than F_{k-2}. We then repeat the process on N_1, to find the next digit of our encoding and the next value N_2. The cycle is continued until N has been thus exhausted. For instance, if $N = 33$, we find $F_8 = 21$ and $N_1 = 12$. Then we subtract $F_6 = 8$ to form $N_2 = 4$, $F_4 = 3$ to form $N_2 = 1$, and $F_2 = 1$, leaving 0. Thus we may write

$$33_{10} \rightarrow (8642)_{\text{Fibonacci encoded}}$$

Just as was the case with our binomial number system, the absence of any simple relation among the digits of our encoding (the Fibonacci index numbers) impedes the

TABLE 7.15 ENCODING OF
INTEGERS BY FIBONACCI
INDICES

N	d_3	d_2	d_1
0	0	0	0
1	0	0	2
2	0	0	3
3	0	0	4
4	0	4	2
5	0	0	5
6	0	5	2
7	0	5	3
8	0	0	6
9	0	6	2
10	0	6	3
11	0	6	4
12	6	4	2
13	0	0	7
14	0	7	2
15	0	7	3
16	0	7	4
17	7	4	2
18	0	7	5
19	7	5	2
20	7	5	3

execution of arithmetic in this number system. The mere fact that an algorithmic procedure exists for encoding numbers into this form and for decoding them into conventional form is of little interest from the standpoint of computer arithmetic. The reader may find it an interesting pastime to examine other mathematical functions that take on integer values and to find ways in which linear combinations of such functions might be chosen in such a way as to represent the integers, at least over some finite range. There are quite possibly number systems yet to be invented that are characterized by ease of encoding and decoding and by simple manipulations equivalent to the ordinary arithmetic operations.

7.11 RATIONAL NUMBER ARITHMETIC

Since the computer is a finite device, it cannot accept as inputs nor produce by its internal operations any irrational numbers, that is, numbers such as π or ϵ that are not the ratio of two integers. The best it can do is to approximate such numbers with a finite string of digits, appropriately rounded. The situation is not necessarily better for rational numbers, because in any base there are some numbers that cannot be expressed with a finite string of digits even though the numerator and the denominator separately may be expressible with very short strings of digits. For example, the number $\frac{1}{3}$ in base 10 requires an infinite sequence of 3's for its representation in radix form, that is, as a weighted sum of powers of the radix, although its numerator and denominator each require only one digit. (Of course, in base 3 this fraction may be written simply as 0.1, but a change of base that accommodates one number will not serve for others.)

One way of avoiding this representational error is to represent each rational number separately by the two integers that form its numerator and denominator, bypassing the implied division with its probably nonterminating result. Arithmetic operations on such numbers are clearly somewhat more complicated than operations on conventional numbers. Let $X = x/x'$ and $Y = y/y'$ be two numbers expressed as numerator–denominator pairs. Then for multiplication we have

$$Z = X \times Y$$

$$= \frac{x}{x'} \times \frac{y}{y'}$$

$$= \frac{z}{z'}$$

where $z = x \times y$ and $z' = x' \times y'$. Similarly, for division

$$Z = \frac{X}{Y}$$

$$= \frac{x \times y'}{x' \times y}$$

$$= \frac{z}{z'}$$

For addition or subtraction

$$Z = X \pm Y$$

$$\frac{z}{z'} = \frac{x}{x'} \pm \frac{y}{y'}$$

$$= \frac{xy' \pm x'y}{x'y'}$$

Thus multiplication and division of numbers in this form require two multiplications, and addition (or subtraction) requires three multiplications and an addition (or subtraction). Because of these multiplications, a succession of operations involving such numbers can produce very large values for numerators and denominators. For instance, if we add $\frac{2}{5} + \frac{2}{5} + \frac{2}{5} + \frac{2}{5} + \frac{2}{5}$ we get

$$\frac{2}{5} + \frac{2}{5} = \frac{10 + 10}{25} = \frac{20}{25}$$

$$\frac{20}{25} + \frac{2}{5} = \frac{100 + 50}{125} = \frac{150}{125}$$

$$\frac{150}{125} + \frac{2}{5} = \frac{750 + 250}{625} = \frac{1000}{625}$$

$$\frac{1000}{625} + \frac{2}{5} = \frac{5000 + 1250}{3125} = \frac{6250}{3125}$$

Of course, we see at once that the numbers would have been much smaller if each fraction had been reduced to lowest terms by dividing out the greatest common divisor (GCD) of numerator and denominator after each addition. It is quite possible to program a computer to carry out the various steps of rational arithmetic, including calculation of the GCD and its cancellation from the numerator and denominator, but such programs require so many steps that they are very slow and inefficient. Furthermore, even with cancellation of the GCD at each step, an extended sequence of computations can result in very large numerators and denominators, which can be of very different sizes.

The difficulties of coping with constantly varying field lengths for the numerator–denominator (N/D) pairs have resulted in proposals for use of fixed-size digit groups. D. W. Matula [43–45] has devised two such modes of representation which he calls the fixed-slash and floating-slash systems. Each one uses a fixed-size group to represent an N/D pair. In the *fixed-slash* format, the sizes of the numerator and denominator fields are equal, but in the *floating-slash* format, the position of the slash

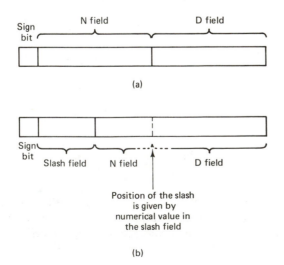

(a)

(b)

Figure 7.5   (a) Fixed-slash and (b) floating-slash formats.

(i.e., the fraction bar) is variable and is indicated by a third field. The formats are shown in Fig. 7.5.

In any fixed-size format, only a finite precision is available for representing numbers, and we are faced with the problem of how best to use this limited precision for representing numbers whose numerator and/or denominator exceed the available size. For instance, if we wish to approximate the fraction $^{371}/_{853} = 0.43493552\ldots$ in a fixed-slash format having only two decimal digits in each field, we could truncate N and D to form $^{37}/_{85} = 0.43529411\ldots$, but this is not as good an approximation as $^{10}/_{23} = 0.43478260\ldots$.

A succession of approximations to a number in N/D form can be generated by writing the number as a continued fraction. Our example of $^{371}/_{853}$ can be written as

$$\frac{371}{853} = \frac{1}{853/371}$$

$$= \cfrac{1}{2 + \cfrac{111}{371}}$$

$$= \cfrac{1}{2 + \cfrac{1}{371/111}}$$

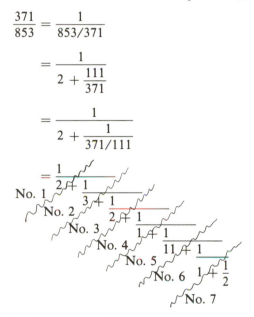

The successive approximations are obtained by truncating the expansion along the successive wavy lines, giving

$$\text{approximation No. 1} = \frac{1}{2} = 0.50\ldots$$

$$\text{approximation No. 2} = \frac{3}{7} = 0.42857142\ldots$$

$$\text{approximation No. 3} = \frac{7}{16} = 0.43750\ldots$$

$$\text{approximation No. 4} = \frac{10}{23} = 0.43478260\ldots$$

$$\text{approximation No. 5} = \frac{117}{269} = 0.43494423\ldots$$

$$\text{approximation No. } 6 = \frac{127}{292} = 0.43493150\ldots$$

$$\text{approximation No. } 7 = \frac{371}{853} = 0.43493552\ldots$$

It is easy to see that, in general, for expansions formed in this way, the successive values form a series that oscillates about the true value, each successive term being closer to the true value of N/D than its predecessor. Thus to get an approximation that is as close as possible to the true value and that fits the fixed-field size constraint, we use the highest-numbered approximation whose N and D lengths fit that constraint. In our example, approximation 4 is the highest-numbered approximation using two decimal digits each for N and for D.

Algorithms for finding the best such approximant have been described in the literature, and rational arithmetic continues to be an interesting research area.

REFERENCES

1. Chen, T. C., and I. T. Ho, Storage-efficient representation of decimal data, *Commun. ACM*, vol. 18, no. 1, Jan. 1975, pp. 49–52.

2. Agrawal, D. P., On arithmetic inter-relationships and hardware interchangeability of mega-binary and binary systems, *Proc. 4th Symp. on Comp. Arith.*, Oct. 1978, IEEE Cat. no. 78CH1412-6C, pp. 88–96.

3. Sankar, P. V., S. Chakrabarti, and E. V. Krishnamurthy, Arithmetic algorithms in a negative base, *IEEE Trans. Comp.*, vol. C-22, no. 2, Feb. 1973, pp. 120–125.

4. Agrawal, D. P., Arithmetic algorithms in a negative base, *IEEE Trans. Comp.*, vol. C-24, no. 10, Oct. 1975, pp. 998–1000.

5. de Regt, M. P., Negative radix arithmetic, *Comp. Design*, vol. 16, May 1967, pp. 52–63.

6. Pawlak, Z., An electronic digital computer based on the "−2" system, *Bull. de l'Acad. Polonaise des Sciences, Série des sci. tech.*, vol. 7, no. 12, 1959, pp. 713–721.

7. Mitchell, J. N., Jr., Computer multiplication and division using binary logarithms, *IRE Trans. El. Comp.*, vol. EC-11, no. 4, Aug. 1962, pp. 512–517.

8 Kingsbury, N. G., and P. J. W. Rayner, Digital filtering using logarithmic arithmetic, *Electron. Lett.*, vol. 7, no. 2, 1971, pp. 56–58.

9. Hall, E. L., D. D. Lynch, and S. J. Dwyer III, Generation of products and quotients using approximate binary logarithms for digital filtering applications, *IEEE Trans. Comp.*, vol. C-19, no. 2, Feb. 1970, pp. 97–105.

10. Combet, M., H. Van Zonneveld, and L. Verbeek, Computation of the base 2 logarithm of binary numbers, *IEEE Trans. El. Comp.*, vol. EC-14, no. 6, Dec. 1965, pp. 863–867.

11. Brubaker, T. A., and J. C. Becker, Multiplication using logarithms implemented with read-only memory, *IEEE Trans. Comp.*, vol. C-24, no. 8, Aug. 1975, pp. 761–765.

12. Swartzlander, E. E., Jr., and A. G. Alexopoulos, The sign/logarithm number system, *IEEE Trans. Comp.*, vol. C-24, no. 12, Dec. 1975, pp. 1238–1242.

13. Lee, S. C., and A. D. Edgar, The focus number system, *IEEE Trans. Comp.*, vol. C-26, no. 11, Nov. 1977, pp. 1167–1170.

14. Swartzlander, E. E., Jr., Comment on "The focus number system," *IEEE Trans. Comp.*, vol. C-28, no. 9, Sept. 1979, p. 693.

15. Lee, S. C., and A. D. Edgar, Addendum to "The focus number system," *IEEE Trans. Comp.*, vol. C-28, no. 9, Sept. 1979, p. 693.

16. Szabo, N., and R. Tanaka, *Residue Arithmetic and Its Applications to Computer Technology*, McGraw-Hill, New York, 1967.

17. Garner, H. L., The residue number system, *IRE Trans. El. Comp.*, vol. EC-8, no. 2, June 1959, pp. 140–147.

18. Rao, T. R. N., and A. K. Trehan, Binary logic for residue arithmetic using magnitude index, *IEEE Trans. Comp.*, vol. C-19, no. 8, Aug. 1970, pp. 752–757.

19. Sasaki, A., Addition and subtraction in the residue number system, *IEEE Trans. El. Comp.*, vol. EC-16, no. 9, Apr. 1967, pp. 157–164.

20. Sasaki, A., The basis for implementation of additive operations in the residue number system, *IEEE Trans. Comp.*, vol. C-17, no. 11, Nov. 1968, pp. 1066–1073.

21. Jullien, G. A., and W. C. Miller, Application of the residue number system to computer processing of digital signals, *Proc. 4th Symp. on Comp. Arith.*, Oct. 1978, IEEE Cat. no. 78CH1412-6C, pp. 220–225.

22. Banerji, D. K., A novel implementation for addition and subtraction in residue number systems, *IEEE Trans. Comp.*, vol. C-23, no. 1, Jan. 1974, pp. 106–109.

23. Banerji, D. K., and J. A. Brzowjowski, Sign detection in residue number systems, *IEEE Trans. Comp.*, vol. C-18, no. 4, Apr. 1969, pp. 313–320.

24. Vinogradov, I. M., *Elements of Number Theory*, Dover, New York, 1954.

25. Jullien, G. A., Residue number scaling and other operations using ROM arrays, *IEEE Trans. Comp.*, vol. C-27, no. 4, Apr. 1978, pp. 325–336.

26. Kinoshita, E., H. Kosako, and Y. Kojima, General division in the symmetric residue number system, *IEEE Trans. Comp.*, vol. C-22, no. 2, Feb. 1973, pp. 134–142.

27. Jullien, G. A., Implementation of multiplication, modulo a prime number, with applications to number theoretic transforms, *IEEE Trans. Comp.*, vol. C-29, no. 10, Oct. 1980, pp. 899–905.

28. Taylor, E. J., and C. H. Huang, An autoscale residue multiplier, *IEEE Trans. Comp.*, vol. C-31, no. 4, Apr. 1982, pp. 321–325.

29. Bachman, G., *Introduction to p-adic Numbers and Valuation Theory*, Academic Press, New York, 1964.

30. Krishnamurthy, E. V., Matrix processors using *p*-adic arithmetic for exact linear computations, *IEEE Trans. Comp.*, vol. C-26, no. 7, July 1977, pp. 633–639.

31. Horspool, R. N., and E. C. R. Hehner, Exact arithmetic using a variable-length *p*-adic representation, *Proc. 4th Symp. on Comp. Arith.*, Oct. 1978, IEEE Cat. no. 78CH1412-6C, pp. 10–14.

32. Avizienis, A., Signed-digit number representations for fast parallel arithmetic, *IRE Trans. El. Comp.*, vol. EC-10, no. 3, Sept. 1961, pp. 389–400.

33. Avizienis, A., Binary-compatible signed-digit arithmetic, *Proc. AFIPS Fall Joint Comp. Conf.*, 1964, pp. 663–671.

34. Tung, C., Signed-digit division using combinational arithmetic nets, *IEEE Trans. Comp.*, vol. C-19, no. 8, Aug. 1970, pp. 746–748.

35. Knuth, D. E., An imaginary number system, *Commun. ACM*, vol. 3, no. 4, Apr. 1960, pp. 245–247.

36. Penney, W., A "binary" system for complex numbers, *Jl. ACM*, vol. 12, no. 2, Apr. 1965, pp. 247–248.

37. Holmes, W. N., Representation for complex numbers, *IBM Jl. R&D*, vol. 22, no. 4, July 1978, pp. 429–430.

38. Yuen, C. K., On the floating-point representation of complex numbers, *IEEE Trans. Comp.*, vol. C-24, no. 8, Aug. 1975, pp. 846–848.

39. Knuth, D. E., *The Art of Computer Programming*, vol. 1, Addison-Wesley, Reading, Mass., 1968, p. 72.

40. Ibid., p. 85.

41. Ibid., pp. 78–86.

42. *The Fibonacci Quarterly*, The Fibonacci Assn., University of Santa Clara, Calif.

43. Matula, D. W., Fixed-slash and floating-slash rational arithmetic, *Proc. 3rd Symp. on Comp. Arith.*, IEEE, Nov. 1975, pp. 90–91.

44. Kornerup, P., and D. W. Matula, A feasibility analysis of fixed-slash rational arithmetic, *Proc. 4th Symp. on Comp. Arith.*, Oct. 1978, IEEE Cat. no. 78CH1412-6C, pp. 39–47.

45. Matula, D. W., and P. Kornerup, A feasibility analysis of binary fixed-slash and floating-slash number systems, ibid., pp. 29–38.

EXERCISES

7.1. Use the rules for mixed-base multiplication to find the area of a rectangle whose two sides are $x = 2$ furlongs, 76 yards, 1 foot 3 inches and $y = 3$ furlongs, 200 yards, 2 feet 9 inches. In what units is your result? Express your result in the form $x \cdot y = d_4 d_3 d_2 d_1 d_0$, where

$d_0 = $ number of elements of unit weight
$d_1 = $ number of elements of weight $= 12$
$d_2 = $ number of elements of weight $= 12 \times 3$
$d_3 = $ number of elements of weight $= 12 \times 3 \times 220$
$d_4 = $ number of elements of weight $= 12 \times 3 \times 220 \times 8$

Make sure that no d_j exceeds the weight associated with its position. Verify your arithmetic by converting x and y to inches, multiplying them, and then expressing the result in the original mixed-base form.

7.2. The Duodecimal Society is a group that is promoting the adoption of the radix 12 as the base of our common number system instead of 10. Although the endeavor is quixotic, it is clear that 12 has some useful advantages because it has 2, 3, 4, and 6 as factors, whereas 10 has only 2 and 5. Show how decimal fractions that have 3 or 6 or 9 as their denominator can be expressed in nonrepeating form if we use base 12. What new repeating fractions occur in base 12 but not in base 10?

7.3. Given the congruence $ax \equiv b \bmod w$, where a and b are constants smaller than w, we may solve for the residue class x by successively finding related congruences with smaller moduli. Thus $ax \equiv b \bmod w$ implies that

$$ax - b = wk_1$$

where k_1 is an as yet unknown integer. Then we must have

$$b + wk_1 = 0 \bmod a$$

which is a new congruence with smaller modulus. We repeat the process until we obtain a congruence whose solution is obvious, and then go back up our chain of $\ldots k_2, k_1$ to solve the given congruence. Use this procedure to find the residue classes x satisfying the following congruences.
(a) $3x \equiv 4 \bmod 17$
(b) $17x \equiv 1 \bmod 23$
(c) $63x \equiv 52 \bmod 97$

7.4. The four moduli 29, 31, 33, and 35 (which are pairwise relatively prime) can be used to define somewhat more than 10^6 numbers and could thus be a useful set in many practical applications.
(a) Given a number $x \equiv 26 \bmod 29$, $x \equiv 24 \bmod 31$, $x \equiv 26 \bmod 33$, and $x \equiv 15 \bmod 35$, determine the value of x.
(b) Find the residue number forms of $x = 9057$ and $y = 193$. Then use these forms to find the residue number forms for their sum and their product.
(c) Note that the product of 193×9057 is beyond the range of this set of moduli. What number that is within range (i.e., less than $29 \times 31 \times 33 \times 35$) has the same residues as this product?

7.5. Consider the digit set $D = \{-1, 0, 1\}$ in conjunction with radix $r = 2$ and radix $r = 3$.
(a) Show an addition table for each radix.
(b) How many numbers are representable in each radix, using this digit set?
(c) What are the largest magnitude positive and negative numbers representable in each radix, using this digit set?
(Note that D is a redundant digit set for $r = 2$ but not for $r = 3$.)

7.6. Using the four different radix 10 redundant signed-digit sets, express the following numbers and carry out the indicated addition in signed-digit form. (Use the form that a totally parallel adder would produce if the conventional form were added to 0.)

(a)	678	**(b)**	291	**(c)**	123
	+ 439		+ 876		+ 234

7.7. The number of numbers expressible in the "binomial number system" having three digits and a maximum value k for d_2 was stated to be

$$\sum_2^k \binom{j}{2} = \binom{k + 1}{3}$$

(a) Prove that this is so.
(b) How many numbers are expressible with four digits and a maximum value of k for d_3?
(c) How many numbers are expressible with n digits and a maximum value of k for d_{n-1}?

7.8. Find the best possible rational approximation to $^{473}/_{684}$, using two decimal digits each in the numerator and in the denominator.

7.9. Using the digit set

$$D = \{0, 1, 20, 21, 40, 41, 60, 61, 80, 81\}$$

with radix $r = 10$, represent the integers 0 through 99. Each of these integers can be expressed in no more than three digits chosen from this set.

8

Computation of Functions

8.1 INTRODUCTION

We have now examined a variety of number systems and ways of using them to carry out the elementary arithmetic operations of addition, subtraction, multiplication, and division. Any computer that can do these operations can do any of the more elaborate functional computations, since they can all be expressed in terms of these elementary arithmetic operations. However, these four operations are not a required set of primitives for a computer, since they in turn can be programmed in terms of yet more elementary operations such as NAND or NOR. As computer technology advances, designers find it increasingly feasible to build into the machine the hardware facilities for performing computations that are much more elaborate than the elementary functions, and we turn now to a consideration of some of them.

8.2 FUNCTION EVALUATION BY ITERATIVE COTRANSFORMATION [1,2]

In engineering systems, it is common practice to employ the principle of feedback to enforce a functional relation between an output variable y and a set X of input variables $\{x_1, x_2, \cdots, x_n\}$. For example, if the input voltage of a high-gain amplifier is forced toward zero by the simultaneous action of an input signal x and a signal $f^{-1}(y)$ fed back from the output voltage y, the output y assumes whatever value is dictated by the constraint toward zero and by the nature of the feedback, so that $y = -f(x)$. As used in digital systems, the technique typically involves two formulas that have some or all of their variables in common. Iterations of one of these formulas, forcing it to a particular value, cause the other one to yield the desired result. We saw an example of

221

this method in Chapter 5, in the multiplicative iteration method of division, in which we formed a succession of factors m_i in such a way that

$$D \times \prod_{i=1}^{n} m_i \rightarrow 1 \qquad \text{as } n \rightarrow \infty$$

thus ensuring that

$$N \times \prod_{i=1}^{n} m_i \rightarrow Q$$

We shall examine now some other applications of this method.

8.2.1 The CORDIC Procedure [3–6]

Among the many functions that can be generated by special-purpose hardware, the trigonometric functions have been of particular interest and were some of the first to be implemented. Computational requirements of the navigation problem were the stimulus for these developments, and the algorithms for CORDIC* that were described by Volder [3] provide us a good starting point for examining the broad class of techniques that have been used not only for the trigonometric functions but for many other transcendental functions as well.

Consider first the problem of rotating a vector (R, β) through a specified positive angle θ. Assume that the original vector is expressed in terms of its rectangular components X and Y and we wish to find the components X' and Y' of the rotated vector, as shown in Fig. 8.1. We see that

$$X' = X \cos \theta - Y \sin \theta \qquad (8.1)$$

and

$$Y' = X \sin \theta + Y \cos \theta \qquad (8.2)$$

Thus

$$\frac{X'}{\cos \theta} = X - Y \tan \theta \qquad (8.3)$$

$$\frac{Y'}{\cos \theta} = Y + X \tan \theta \qquad (8.4)$$

The CORDIC algorithm is an iterative procedure in which each step i rotates the vector in one or the other direction through an angle of magnitude $\alpha_i = \tan^{-1} 2^{-i}$. The direction ($+$ or $-$) for each α_i is chosen as $+$ if $(\theta - \sum_{0}^{i=1} \alpha_j) > 0$ and as $-$ if $(\theta - \sum_{0}^{i=1} \alpha_j) < 0$. Thus, if we let θ_i denote the total angle through which the vector has been rotated by all steps through step i, $\theta_i = \pm \alpha_0 \pm \alpha_1 \pm \alpha_2 \cdots \pm \alpha_i$. This series converges for all angles of magnitude less than about 100°, that is, 1.74 radians. Since

---

Coordinate Rotational Digital Computer.

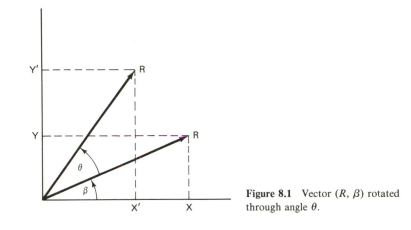

Figure 8.1 Vector (R, β) rotated through angle θ.

any angle $|\theta| > 90°$ can be expressed in terms of a related angle of magnitude less than 90°, we see that the convergence of the iterative process can be assured for any angle if it is first reduced to one less than 90° in magnitude.

The iterative rotation through $\pm\alpha_i$ is used to compute X' and Y' iteratively by means of

$$X_{i+1} = X_i - Y_i \tan \alpha_i = X_i \mp Y_i \times 2^{-i} \tag{8.5}$$

$$Y_{i+1} = Y_i + X_i \tan \alpha_i = Y_i \pm X_i \times 2^{-i} \tag{8.6}$$

where $i = 0, 1, 2, \ldots$, and X_0 and Y_0 are, respectively, the original components X and Y. As we see from (8.3) and (8.4), the X_{i+1} and Y_{i+1} computed from (8.5) and (8.6) are components of a vector that is longer than the vector (X_i, Y_i) in the ratio $1/\cos \alpha_i$. Continuation of the iterations (8.5) and (8.6) until θ_i approximates θ therefore produces X and Y components that are K_c times as large as the true X' and Y', where K_c is given by

$$K_c = \frac{1}{\cos \alpha_0} \cdot \frac{1}{\cos \alpha_1} \cdot \frac{1}{\cos \alpha_2} \cdots$$

$$\simeq \sqrt{1 + 2^{-0}} \times \sqrt{1 + 2^{-2}} \times \sqrt{1 + 2^{-4}} \times \cdots$$

$$= 1.646760255 \ldots$$

which is Volder's value through $i = 24$. Division by this factor forms the true X' and Y' values from the X and Y values formed by repeated applications of (8.5) and (8.6).

Table 8.1 shows a few steps of the rotation of the vector $(X = 1, Y = 1)$ through an angle of $-30°$. For convenience, the angles are given in degrees, although in the computer registers, each angle would be expressed as a binary fraction of π radians, to whatever precision is desired. (If angles of rotation exceeding 90° are to be handled, a first step that rotates the vector by a suitable multiple of $\pm90°$ precedes the iterations.

TABLE 8.1

i	$\tan \alpha_i = 2^{-i}$	α_i	θ_i	$\theta - \theta_i$	X_{i+1}	Y_{i+1}
				$-30°$	1	1
0	1	$-45°$	$-45°$	$15°$	2	0
1	2^{-1}	$+26.6°$	$-18.4°$	$-11.6°$	2	1
2	2^{-2}	$-14.04°$	$-32.44°$	$+2.44°$	2.25	0.5
3	2^{-3}	$+7.13°$	$-25.31°$	$-4.69°$	2.1875	0.7813
4	2^{-4}	$-3.58°$	$-28.89°$	$-1.11°$	2.236	0.6445
5	2^{-5}	$-1.79°$	$-30.68°$	$+0.68°$	2.256	0.574
6	2^{-6}	$+0.90°$	$-29.78°$	$-0.22°$	2.247	0.609
7	2^{-7}	$-0.45°$	$-30.23°$	$+0.23°$	2.252	0.591
8	2^{-8}	$+0.22°$	$-30.01°$	$+0.01°$	2.250	0.600

That preliminary step has not been included in Table 8.1.) We would expect our final values for X and Y to be

$$X = \sqrt{2} \cos 15° \times \prod_{i=0}^{n} \frac{1}{\cos \alpha_i} \simeq 2.24951634 \ldots$$

$$Y = \sqrt{2} \sin 15° \times \prod_{i=0}^{n} \frac{1}{\cos \alpha_i} \simeq 0.60275609 \ldots$$

The limited precision of Table 8.1 compares well with these values. The true magnitudes of the components, of course, are obtained by dividing these numbers by the magnifying factor K_c. (Alternatively, we could have divided the initial X and Y values by K_c.) Note that the hardware requirements for performing these steps are quite simple. We need a register for X_i and one for Y_i, shifting means for each (to do the multiplication by 2^{-i}), and two adders. We also need to accumulate $\theta - \theta_i$ and to test its sign to determine the sign of the next step. Figure 8.2 shows a block diagram of these elements. The crossover paths A and B carry the shifted (and possibly complemented) values of X_i and Y_i to be added or subtracted as specified by (8.5) and (8.6). The sign of α_i is determined by the sign bit of the $(\theta - \theta_i)$ register, and the various values of α_i must be provided on successive steps. These values are typically stored in a read-only memory. The details of the circuitry for shifting are heavily technology dependent, and both shifting matrices [4] and a split register configuration [6] have been used. The shifting circuitry may easily constitute a major part of the required circuitry for implementing the algorithm.

The structure and the algorithm can be easily adapted to several other purposes besides the rotation described above. In the following examples, we have assumed in each case that $0° < \theta < 90°$, so that X and Y components are both positive. The procedures are easily adapted to vectors in the second, third, or fourth quadrants of the (x, y) plane by use of well-known trigonometric identities, so that no rotation of more than $90°$ is required.

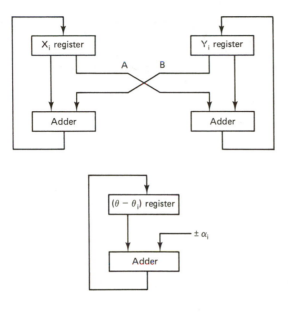

Figure 8.2   Logic structure for CORDIC algorithm.

1. Given a vector (R, θ), we find its components X' and Y' by setting $X_0 = R$, $Y_0 = 0$, and rotating through θ.

2. Given an angle θ, we find $\sin \theta$ and $\cos \theta$ by setting $X_0 = 1$, $Y_0 = 0$, and rotating through θ, to form $X' = \cos \theta$ and $Y' = \sin \theta$ (after multiplying by $1/K_c$).

3. Given an angle θ, we find $\tan \theta$ by forming $\sin \theta$ and $\cos \theta$, as in (8.2), and then forming their quotient.

4. Given the components X and Y, we find (R, θ) by choosing successive α_i so that $Y_{i+1} \rightarrow 0$ as the rotation continues. (If $Y_i > 0$, sign of α_i is negative, and vice versa.) Then $R = X_{i+1} \times 1/\!/K_c$) and $\theta = \theta_i$, the accumulated sum of $\pm\alpha_i$.

5. Given X_1 and Y_1, to find $\theta = \tan^{-1}(Y_1/X_1)$, set $X = X_1$ and $Y = Y_1$. Rotate the vector by choosing signs for successive α_i so that $Y_{i+1} \rightarrow 0$. Then $\theta = \theta_i$.

Note that in using this algorithm, we have implicitly assumed that every computation will use the same number of iterations, so that a common value of magnifying factor K_c can be used. It may happen, of course, that fewer steps will suffice in some particular instance, and the algorithm could then be terminated earlier. To do this, however, would require that we also compute at each step $K_{i+1} = K_i \times (1/\cos\alpha_i)$ so that the appropriate correction factor could be applied in the event of early termination. The more common practice is to carry out a fixed number of iterations.

8.2.2 Extensions of CORDIC

Following publication of the CORDIC algorithms, there have been several extensions of that technique for the computation of related functions [1, 7–10]. Since the trigono-

metric functions, the hyperbolic functions, and simple exponentials are all closely related, it is not surprising that similar algorithms may be used to compute all of them.

The close relation between the trigonometric functions and the hyperbolic functions is evident when we examine their defining geometries. The trigonometric functions ("circular" functions) are defined with respect to the circle $x^2 + y^2 = R^2$, as
shown in Fig. 8.3. By definition:

$$\cos \beta = \frac{X_0}{R}$$

$$\sin \beta = \frac{Y_0}{R} \qquad (8.7)$$

$$\tan \beta = \frac{Y_0}{X_0}$$

and the angle β is defined as being $2/R^2 \times$ the area included between the x axis, the
circle of radius R, and the vector V from the origin to (X_0, Y_0). For $R = 1$, the cosine
and the sine become simply X_0 and Y_0, respectively.

To define the hyperbolic functions, we use the hyperbola $x^2 - y^2 = R^2$, as
shown in Fig. 8.4. By definition

$$\cosh \beta = \frac{X_0}{R}$$

$$\sinh \beta = \frac{Y_0}{R} \qquad (8.8)$$

$$\tanh \beta = \frac{Y_0}{X_0}$$

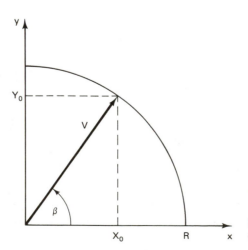

Figure 8.3 Defining geometry for the
circular functions.

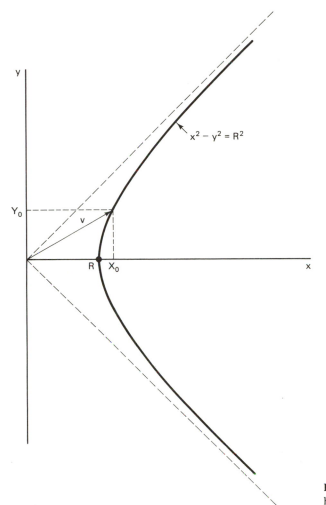

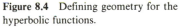

Figure 8.4 Defining geometry for the hyperbolic functions.

and the hyperbolic angle β is defined as being $2/R^2 \times$ the area included between the x axis, the hyperbola of "radius" R, and the vector V from the origin to (X_0, Y_0). For $R = 1$, $\cosh \beta$ and $\sinh \beta$ are simply X_0 and Y_0, respectively.

Now consider the rotation of vector $V = (X_0, Y_0)$ to a new position $V = (X', Y')$, where the (hyperbolic) angle of rotation is θ and where we take $R = 1$ for convenience. We find

$$X' = \cosh (\beta + \theta) = \cosh \beta \cosh \theta + \sinh \beta \sinh \theta$$
$$Y' = \sinh (\beta + \theta) = \sinh \beta \cosh \theta + \cosh \beta \sinh \theta$$
$$X' = X_0 \cosh \theta + Y_0 \sinh \theta$$
$$Y' = Y_0 \cosh \theta + X_0 \sinh \theta$$

(8.9)

We may write

$$\frac{X'}{\cosh \theta} = X_0 + Y_0 \tanh \theta$$

$$\frac{Y'}{\cosh \theta} = Y_0 + X_0 \tanh \theta$$

(8.10)

Now we take θ to be

$$\theta = \pm \, \alpha_1 \pm \alpha_2 \pm \cdots$$

$$\theta_i = \sum_1^i \pm \, \alpha_j$$

(8.11)

where $\alpha_j = \tanh^{-1} 2^{-j}$.* This leads to

$$X_i = X_{i-1} \pm Y_{i-1} \times 2^{-i}$$

$$Y_i = Y_{i-1} \pm X_{i-1} \times 2^{-i}$$

Just as in the trigonometric case, we see that X_i and Y_i computed from this pair of equations are components that differ in length from X_i' and Y_i' in the ratio $X_i = X_i'/\cosh \alpha_i$ and $Y_i = Y_i'/\cosh \alpha_i$. Continuation of the iterations until θ_i approximates θ therefore produces X and Y components that are K_H times the true X' and Y', where K_H is given by

$$K_H = \frac{1}{\cosh \alpha_1} \cdot \frac{1}{\cosh \alpha_2} \cdots \simeq 0.828 \ldots$$

To evaluate various quantities related to the hyperbolic functions, we may use procedures closely analogous to the CORDIC algorithms for the circular functions. As with the circular functions, the attractiveness of these procedures lies in the simplicity of the hardware, which needs only shifting and adding circuits to carry out the iteration steps and an accumulating adder whose sign determines the sign of the increment. As in the trigonometric case, a read-only memory provides the values of α_i.

The following are brief descriptions of some of the procedures:

1. Given θ, to find $\sinh \theta$ and $\cosh \theta$, set $X_0 = 1/K_H$ and $Y_0 = 0$. Rotate through θ (by successive increments α_i) to form $X_i = \cosh \theta$ and $Y_i = \sinh \theta$.
2. Given θ, to find $\tanh \theta$, proceed as in item 1 and form $\tanh \theta = \sinh \theta/\cosh \theta$.

---

*However, unlike the series for the circular functions, this series does not converge for all arguments that are of interest to us. In fact, Walther [7] has shown that this simple sequence must be supplemented by including certain of its terms twice. The terms to be repeated are $\alpha_4, \alpha_{13}, \cdots, \alpha_k, \alpha_{3k+1}$. When those terms are included, the series converges for hyperbolic angles less than approximately 1.12. For larger arguments, a preliminary scale adjustment is needed before the shift-and-add iterative process can be used.

3. Given $Z = \sinh \theta$, to find $\theta = \sinh^{-1} Z$, set $X_0 = 1/K_H$ and $Y_0 = 0$. Rotate the vector by choosing successive increments α_i so that $Y_i \to Z$. Then $\theta = \Sigma\, a_i$.

4. Given $Z = \cosh \theta$, to find $\theta = \cosh^{-1} Z$, set $X_0 = 1/K_H$ and $Y_0 = 0$. Rotate the vector by choosing successive α_i so that $Y_i \to Z$. Then $\theta = \Sigma\, \alpha_i$.

5. Given $Y/X = \tanh \theta$, to find $\theta = \tanh^{-1}(Y/X)$, set $X_0 = X$ and $Y_0 = Y$. Rotate the vector by choosing successive increments α_i so that $Y_i \to 0$. The $\theta = \Sigma\, \alpha_i$.

6. Given θ, to find ϵ^θ, follow procedure 1 and then form $\epsilon^\theta = \cosh \theta + \sinh \theta$.

7. Given $Z > 1$, to find $\ln Z$, form $d = (Z - 1)\,/\,(Z + 1)$ so that $Z = (1 + d)\,/\,(1 - d)$. Using the identity $\ln (1 + d)\,/\,(1 - d) = 2 \tanh^{-1}d$, form $\tanh^{-1} d$ as in procedure 5.

As an example of these procedures, Table 8.2 lists successive values in the calculation of $\cosh \theta$ and $\sinh \theta$, for $\theta = 1$, starting from $X_0 = 1/K_H \simeq 1.208$ and $Y_0 = 0$. Note the inclusion of $i = 4$ two times, as specified by Walther. The final values, with the limited precision of the table, are reasonably close to the correct values $\cosh 1 = 1.543$ and $\sinh 1 = 1.175$.

It is possible to evaluate square roots by use of the hyperbolic geometry $X^2 - Y^2 = R^2$. To find $W^{1/2}$, let $X_0 = W + (1/4)$ and $Y_0 = W - 1/4$ so that $W = X^2 - Y^2$. Since any point in the (X, Y) plane that lies between the bounding asymptotes $Y = \pm X$ is on some one of the hyperbolas, the value W equals R^2 for that particular hyperbola, that is, $W^{1/2} = R$. To find R, we rotate the vector $V = (X_0, Y_0)$ so as to force its Y component to zero, which simultaneously forces the X component to R. In principle, the method works for any value W, but in practice, our simple shift-and-add algorithm, using steps $\alpha_i = \tanh^{-1} 2^{-i}$ is subject once again to convergence limitations. Since this series converges for hyperbolic angles no greater than 1.12, it follows that we must limit W to values of 2.35 or less. [Note that $\tanh 1.12 = 0.80757 = Y_0/X_0 = (W - 1/4)/(W + 1/4)$ which leads to $W = 2.35$.] Larger values must be prescaled to reduce them to a size for which the procedure converges.

TABLE 8.2

i	$\tanh \alpha_i = 2^{-i}$	α_i	θ_i	$\theta - \theta_i$	X_i	Y_i
1	0.5	0.55	0.55	0.45	1.208	0.604
2	0.25	0.255	0.805	0.185	1.359	0.906
3	0.125	0.125	0.930	0.070	1.473	1.076
4	0.0625	0.0625	0.993	0.007	1.540	1.168
4	0.0625	0.0625	1.056	−0.056	1.613	1.264
5	0.03125	−0.03125	1.025	−0.025	1.573	1.214
6	0.015625	−0.015625	1.009	−0.009	1.554	1.189
7	0.0078125	−0.0078125	1.001	−0.001	1.545	1.176

8.3 THE FAST FOURIER TRANSFORM [11–12]

No discussion of the computation of functions would be complete without mention of the very important fast Fourier transform (FFT), which has enormously enhanced the capabilities of digital signal processors since its publication in 1965. There is by now an extensive literature on the FFT and its applications and implementations (see, e.g., the bibliography in Ref. 12). We will not attempt to summarize that literature but will limit our discussion to the computational aspects.

The FFT is an algorithm for evaluating the discrete Fourier transform (DFT) in substantially fewer steps than are required by direct evaluation. The DFT is defined by

$$A_k = \sum_{r=0}^{N-1} x_r \times \epsilon^{-j(2\pi/N)rk} \qquad k = 0, 1, \ldots, (N-1)$$

where the quantities x_r are typically the digitized values of a time function at a set of $N = 2^n$ equispaced instants. Direct computation of the N values of A_k requires N multiplications and $(N-1)$ additions for each, a total number of operations of order N^2. It can be shown, however, that the computations can be performed by computing the DFTs of two sequences, one consisting of the $N/2$ even-numbered points and the other of the $N/2$ odd-numbered points. Each of these in turn can be similarly calculated from further sub-sequences, and so on to the lowest level, at which we compute transforms of pairs of points. The computational burden is then reduced to the order of $N \log N$, a very dramatic improvement for such typical values of N as 1024, 2048, or 4096.

We start by setting $W_N = \epsilon^{-j(2\pi/N)}$, so that

$$A_k = \sum_{r=0}^{N-1} x_r \times W_N^{rk} \tag{8.12}$$

The even- and odd-numbered indices r may be separated to give

$$A_k = \sum_{r=0}^{(N/2)-1} x_{2r} \times W_N^{2rk} + \sum_{r=0}^{(N/2)-1} x_{2r+1} \times W_N^{(2r+1)k} \tag{8.13}$$

where the values $2r$ are the even-numbered points in (8.12) and the values $(2r+1)$ are the odd-numbered points.

We note that

$$W_N^{2k} = \epsilon^{-j(2\pi/N)\times 2k}$$

$$= \epsilon^{-j[2\pi/(N/2)]\times 1k} = W_{N/2}^k$$

Therefore (8.13) may be rewritten as

$$A_k = \underbrace{\sum_{r=0}^{(N/2)-1} x_{2r} \times W_{N/2}^{rk}}_{\substack{A_e = \text{DFT over the } N/2 \\ \text{even-numbered} \\ \text{points}}} + W_N^k \underbrace{\sum_{r=0}^{(N/2)-1} x_{2r+1} \times W_{N/2}^{rk}}_{\substack{A_o = \text{DFT over the } N/2 \\ \text{odd-numbered} \\ \text{points}}}$$

Now, because $W_N^k = -W_N^{k+(N/2)}$, we may write separately the values of A_k for $k = 0, 1,$ $\ldots, [(N/2) - 1]$ and for $k = N/2, (N/2) + 1, \ldots, (N - 1)$, as follows:

$$\left.\begin{array}{l} A_k = \mathbf{A}_e(k) + W_N^k A_o(k) \\[2mm] A_{k+(N/2)} = \mathbf{A}_e(k) - W_N^k A_o(k) \end{array}\right\} k = 0, 1, \ldots, \left(\frac{N}{2} - 1\right) \qquad (8.14)$$

where $A_e(k)$ and $A_o(k)$ are, respectively, the DFTs of the even-numbered points and the odd-numbered points, as noted above. Equation (8.14) is important, because it allows us to calculate the A_e and A_o elements in similar fashion from their own even- and odd-numbered subsequences A_{ee}, A_{eo}, A_{oe}, and A_{oo}.

The basic computations involved in (8.14) may be shown as a signal flow graph (Fig. 8.5), whose shape gives the name "butterfly" to the set of operations. Joining arrows represent the addition operation, the sense (+ or −) of each operand being indicated beside its arrowhead. Multiplication is indicated by ×, and signals (numerical values) flow from left to right in time.

A complete signal flow graph for an eight-point FFT is shown in Fig. 8.6. This graph exhibits an important property of the FFT algorithm: it is an "in place" algorithm, since the results of each butterfly computation can be stored in the same place where its original data were. Thus no intermediate storage is required, and the storage provided for the original data suffices for the entire computation.

The ordering of the inputs in their registers derives from successive division of the points into even- and odd-numbered subsets. For instance, x_0 and x_4 are the two even-numbered members of the even-numbered subset (0, 2, 4, 6) of the full set of eight points. This ordering must be used for the inputs if the final outputs are to be found in their correct sequence in the set of registers and if no temporary storage is used for reordering the values after each pass.

If, on the other hand, the input points are stored in the set of registers in their natural order, then the outputs will be found in the sequence corresponding to subsets of subsets, and so on, of even- and odd-numbered points, as shown in Fig. 8.7. It is clear that either procedure involves a reordering if both the input and the output sequences are to appear in the order of their indices.

The computational relief afforded by this algorithm is made clear from the signal flow graphs of Figs. 8.6 and 8.7. It is evident that in the general case of $N = 2^n$,

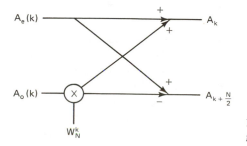

Figure 8.5 "Butterfly" signal flow graph.

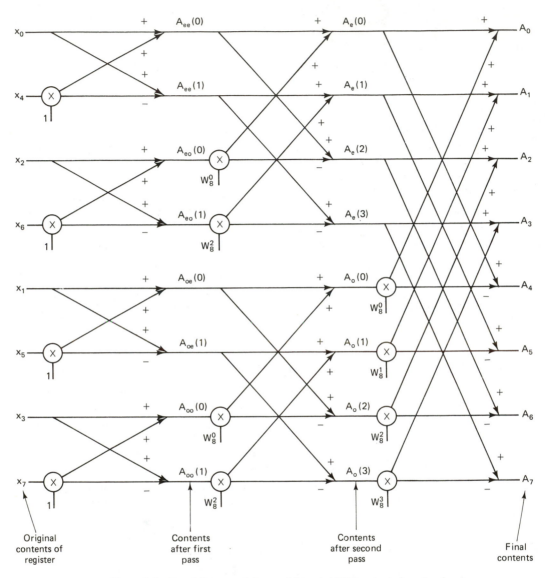

Figure 8.6 Signal flow graph for an eight-point FFT, outputs in natural order.

there will be n columns (or stages) of computations and that each stage will consist of 2^{n-1} butterflies. Thus there will be $N/2 \log_2 N$ multiplications and $N \log_2 N$ additions. Each of these arithmetic operations involves a pair of complex numbers.

(This successive decomposition of the time series into alternating sub-sequences has come to be known by the awkward misnomer *decimation in time*. The term is an unfortunate bit of jargon, since partitioning into two sets is certainly not implied by any word derived from the Latin *decimus*, meaning ten. Although "to decimate" orig-

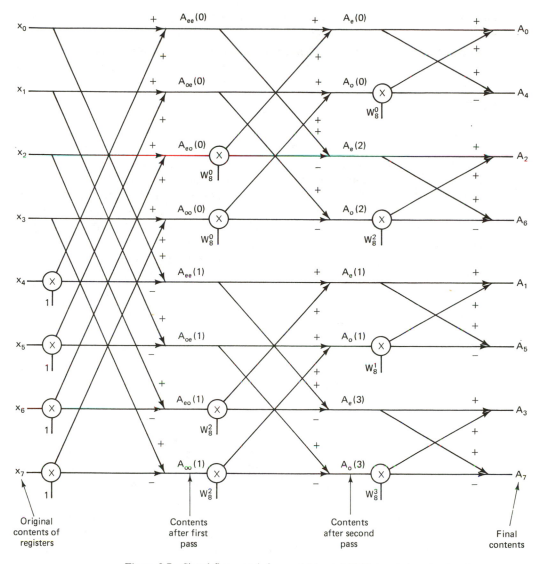

Figure 8.7 Signal flow graph for an eight-point FFT, inputs in natural order.

inally meant "to kill every tenth soldier," its present-day usage means, more loosely, destroying a large part of a group. Computer people may be charged with decimating the English language.)

An alternative decomposition is based on a different subdividing of N points into two sets, one being the first $N/2$ points and the other the second $N/2$, with these sets in turn being similarly subdivided. This procedure leads to a formulation of the algorithm in which even-numbered results A_{2r} and odd-numbered results A_{2r+1} are calculated separately, with the even- and odd-numbered sets being further partitioned into even and odd numbered subsets, and so on. Since it is now the output values ("frequencies") that are treated in odd–even pairs, this form of the algorithm is known (of course!) as *decimation in frequency*. The computational burden is the same for both forms, and both are "in-place" algorithms. Details of this form of the algorithm may be found in Ref. 12.

Since the fast Fourier transform is an $O(N \log N)$ process whereas direct evaluation is an $O(N^2)$ process, the speed advantage of the FFT can be very great, even for computers that are not specially structured to facilitate the FFT. However, it is common practice to use special-purpose attachments known as *array processors* [13] that have their own high-speed arithmetic unit and a memory unit organized to maintain the flow of operands at the speed of the arithmetic unit. Several such systems, designed for use with various popular computer systems, are commercially available.

The regularity of the pattern of computations, as suggested by the signal flow graphs of Figs. 8.6 and 8.7, has made evident many ways in which parallelism can be employed to yield still faster structures for executing the FFT, and many such methods have been described in the literature [14–19].

8.4　FUNCTION EVALUATION USING THE MULTIPLY-ADD ELEMENT

In many computer algorithms, a multiplication operation is followed by addition of some number to the product; that is, it is frequently necessary to form

$$W = X \times Y + Z$$

where X, Y, and Z are n-digit numbers. Of course, the arithmetic-logical unit (ALU) of a general-purpose computer can be programmed to execute this sequence of operations, but it is an easy matter to design logic circuits to accept X, Y, and Z inputs and form the W output in less time than the sequential operations would take if executed under program control.

If such an element is designed for fixed-point numbers, we must decide whether our numbers will be integers or fractions (or perhaps mixed numbers), since this circumstance determines whether Z should be added at the left end of $X \times Y$ (numbers taken as fractions) or at the right end of $X \times Y$ (numbers taken as integers). It is easy to see that if the numbers are integers, the quantity W is always expressible in $2n$ digits; that is, $X \times Y + Z$ cannot overflow the capacity of a double-length register. However, the Z component makes no contributions to the n more significant digits of the double-length result.

If on the other hand, the operands are all taken to be fractions, $X \times Y + Z$ can very easily overflow the register capacity, since Z is now being added at the left end of $X \times Y$.

Some of the more interesting uses of the multiply-add element are in structures that use interconnections of several such elements, and to avoid problems of digit significance or overflow, we might prefer to use MA (multiply-add) elements that execute floating-point arithmetic, although such elements are very complex.

Before considering the design of an MA element, we examine some of the application areas where such devices can be used.

8.4.1 Polynomial Evaluation

A problem that is particularly well suited to computation using MA elements is the evaluation of polynomials. There are several possible ways of interconnecting MA elements to do this evaluation, the most obvious being based on *Horner's method*, in which the polynomial

$$P(x) = a_n x^n + a_{n-1} x^{n-1} \cdots a_1 x + a_0 \tag{8.15}$$

is expressed in nested form as

$$P(x) = (((a_n x + a_{n-1}) \times x + a_{n-2}) \times x + \cdots) \times x + a_0 \tag{8.16}$$

This is an iterative procedure in which we form successively, for $j = 1$ to n,

$$p_j(x) = [p_{j-1}(x)] \times x + a_{n-j} \tag{8.17}$$

with $$p_0(x) = a_n$$

and $$P(x) = p_n(x)$$

The network of MA elements shown in Fig. 8.8 is a direct implementation of this method. Each MA element accepts two n-digit vectors on the product inputs at its left and an n-digit summand on the additive input at the top. It forms a $2n$-digit output at its right after a time delay of T seconds. Thus the final output is produced at the right of the chain after a time delay of nT seconds. Since only one element is active at any instant, this structure is not an efficient one, but it is easily used in pipeline fashion, with all units active and n polynomials being simultaneously computed, one result being formed at the end of the chain in each T-second interval.

If pipelining is not needed, one MA unit used repetitively suffices for evaluation of a single nth-degree polynomial in nT seconds, as shown in Fig. 8.9. One product input receives the value of x at each clock time, and the other receives a_n at the first time and the fed-back outputs at successive times, while the additive input receives in turn $a_{n-1}, a_{n-2}, \cdots$ as required by the nesting form of the polynomial.

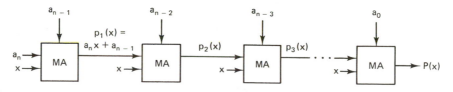

Figure 8.8   Polynomial evaluation with iterated MA elements.

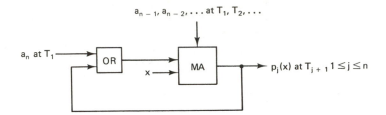

Figure 8.9 Polynomial evaluation with a single MA element.

A structure that computes the polynomial in $O(\log n)$ time units was first described by Estrin [20] and is shown in Fig. 8.10 for the polynomial

$$P(x) = a_7 x^7 + a_6 x^6 + \cdots a_1 x + a_0$$

With n levels, this tree structure can evaluate polynomials of degree $\leq 2n$ in n time units. This structure makes more effective use of the MA elements than does the structure of Fig. 8.8, since more of them are active at each step. Note that by interposing latches between the levels, we can pipeline this structure, too, although the only advantage of the pipelined version of Fig. 8.10 over the pipelined version of Fig. 8.8 is the shorter "transport" delay.

Evaluation of derivatives can be carried out together with the polynomial evaluations by structures that derive some of their inputs from the stages of the polynomial evaluator. The sequential structure of Fig. 8.8 is easily adapted for this purpose, as shown in Fig. 8.11, which displays a structure to evaluate a fourth degree polynomial and its first two derivatives. The structure of Fig. 8.9 has a corresponding form in Fig.

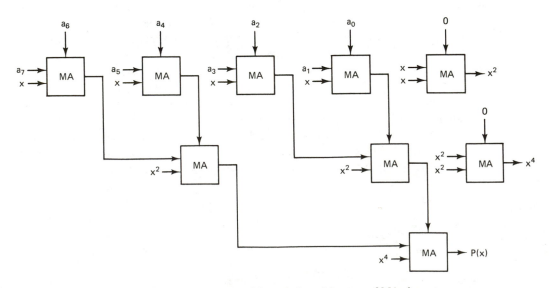

Figure 8.10 Polynomial evauation with a tree of MA elements.

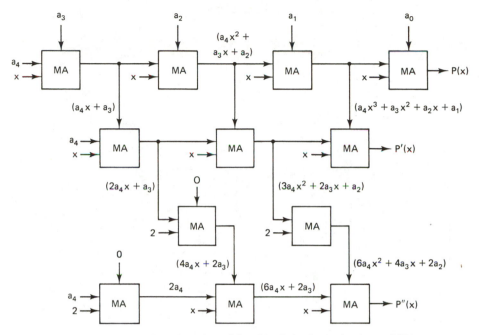

Figure 8.11   Evaluation of a polynomial and its derivatives, using iterated MA elements.

8.12, where a delay element D (equal to the delay time of an MA element) has been included in some loops to keep the various loops in synchronism because of the multiplications by 2, 3, 4, ... that must be incorporated in the loops for the higher derivatives. (The necessary delays or latches for synchronizing have not been included in Fig. 8.12.)

Another interesting application of the polynomial evaluation structure follows from the observation that the general nested multiply-add expression can also be evaluated by either the sequential structure (Figs. 8.8 and 8.9) or the tree structure (Fig. 8.10). Thus the function

$$f = (\cdots ((a_1b_1 + b_2)a_2 + b_3)a_3 + b_4)a_4 + b_5) \cdots + b_n \qquad (8.18)$$

can be expanded to

$$f = a_1a_2a_3 \cdots a_{n-1}b_1 + a_2a_3 \cdots a_{n-1}b_2 + \\ a_3a_4 \cdots a_{n-1}b_3 + \cdots + b_n \qquad (8.19)$$

which is easily evaluated by means of a tree of MA elements. For example, let

$$f = ((a_1b_1 + b_2)a_2 + b_3)a_3 + b_4$$

$$= a_1a_2a_3b_1 + a_2a_3b_2 + a_3b_3 + b_4 \qquad (8.20)$$

Figure 8.13 follows at once.

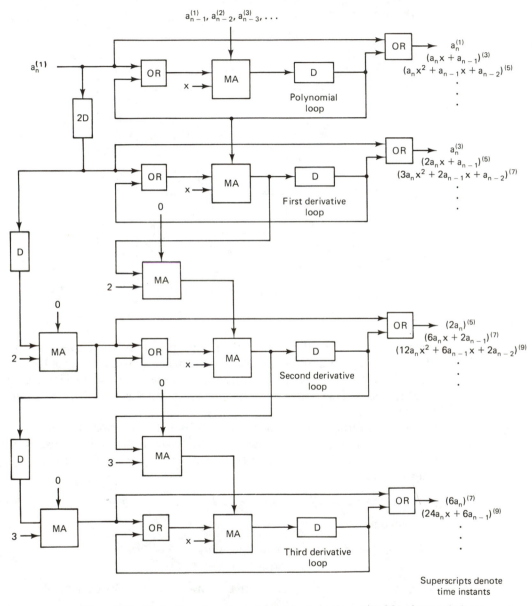

Figure 8.12 Evaluation of a polynomial and its derivatives using MA elements in loops.

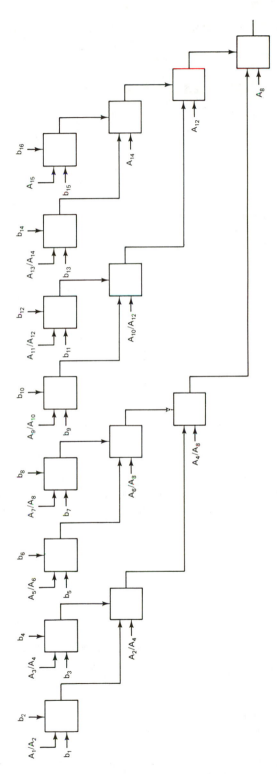

Figure 8.14 Evaluation of $\sum_{i}^{16} A_i b_i$.

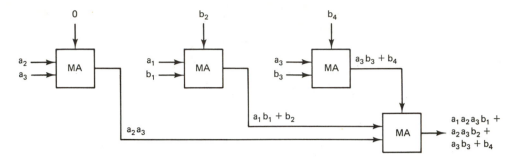

Figure 8.13   Evaluation of the function $f = a_1a_2a_3b_1 + a_2a_3b_2 + a_3b_3 + b_4$.

Now let

$$A_1 = a_1a_2a_3 \cdots a_{n-1}$$

$$A_2 = a_2a_3 \cdots a_{n-1}$$

$$A_3 = a_3a_4 \cdots a_{n-1}$$

$$A_{n-1} = a_{n-1}$$

$$A_n = 1 \tag{821}$$

Our general nested multiply-add expression may then be written

$$f = A_1b_1 + A_2b_2 + \cdots = \sum_{j=1}^{n} A_jb_j \tag{8.22}$$

Thus we see that a vector inner product may be evaluated by a tree of **MA** elements in $O(\log n)$ steps if we know the ratios of the coefficients A_j. Finding these ratios by carrying out the divisions is probably too time consuming to make this method generally useful, but if the ratios are already known, this structure has a significant time advantage. In general, to evaluate

$$\sum_{1}^{2^n} A_jb_j$$

by this method, we need to know K ratios of coefficients, where

$$K = 2^n - (n + 1)$$

for example, for $\sum_{1}^{16} A_jb_j$, we need A_1/A_2, A_3/A_4, A_5/A_6, A_7/A_8, A_9/A_{10}, A_{11}/A_{12}, A_{13}/A_{14}, A_2/A_4, A_6/A_8, A_{10}/A_{12}, and A_4/A_8, as shown in Fig. 8.14.

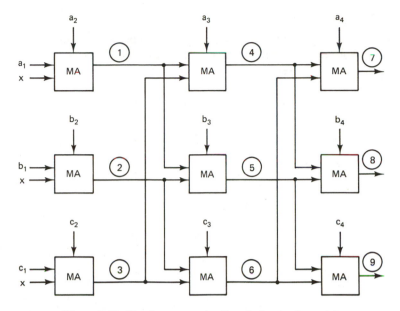

Figure 8.15   Simultaneous evaluation of related polynomials.

As a final example of a network for computing polynomials, we exhibit a structure for simultaneously computing a set of interrelated polynomials, as shown in Fig. 8.15. The following outputs are produced at the numbered points:

1. $a_1 x + a_2$
2. $b_1 x + b_2$ \} First-order polynomials
3. $c_1 x + c_2$

4. $a_3 + (a_1 x + a_2)(c_1 x + c_2)$
$\quad = a_1 c_1 x^2 + (a_2 c_1 + a_1 c_2) x + a_2 c_2 + a_3$
5. $b_3 + (b_1 x + b_2)(a_1 x + a_2)$
$\quad = a_1 b_1 x^2 + (a_1 b_2 + a_2 b_1) x + a_2 b_2 + b_3$
6. $c_3 + (c_1 x + c_2)(b_1 x + b_2)$
$\quad = b_1 c_1 x^2 + (b_1 c_2 + b_2 c_1) x + b_2 c_2 + c_3$

\} Second-order polynomials

7.
8. \} Fourth-order polynomials
9.
etc.

We note that:

1. The chained circuit can be replaced by a closed-loop circuit with no loss in computation speed.

2. The computation time for nth-degree polynomials is $O(\log n)$.
3. The sets of polynomials generated are not independent. That is, we cannot start from a set of independent polynomials and find values of a, b, c, ... that will enable them to be generated in this structure.

8.4.2 Evaluation of Linear Recursions

Consider the linear recursion system $R < n, 3 >$:

$$x_1 = c_1$$

$$x_2 = c_2 + a_{21}x_1$$

$$x_3 = c_3 + a_{31}x_1 + a_{32}x_2$$

$$x_4 = c_4 + a_{41}x_1 + a_{42}x_2 + a_{43}x_3$$

$$x_5 = c_5 \qquad\quad + a_{52}x_2 + a_{53}x_3 + a_{54}x_4 \qquad\qquad (8.23)$$

etc.

Since these equations are sums of products, we can use the MA element in networks for determining in turn the values of $x_1, x_2, \ldots$. As soon as x_j is available, it can be used in the following time interval to compute the three products $a_{(j+1),j} \times x_j$, $a_{(j+2),j} \times x_j$, and $a_{(j+3),j} \times x_j$, which are then added to the results of previous steps in rows j, $j + 1$, and $j + 2$, as shown in Fig. 8.16.

The closed-loop form of this circuit appears in Fig. 8.17. An x_j value is produced

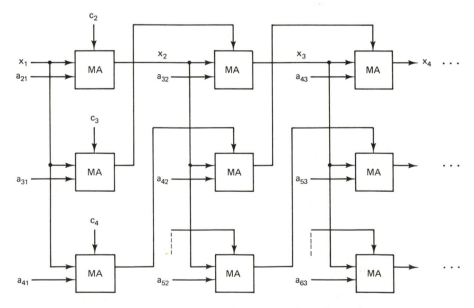

Figure 8.16   Iterated MA elements to evaluate $R < n, 3 >$.

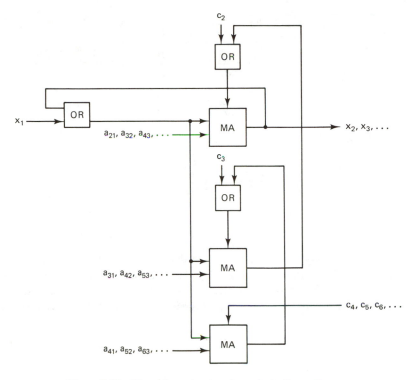

Figure 8.17   Closed loop structure to evaluate $R <n, 3>$.

in each **MA** time unit. In general, it is evident that we can evaluate a recursion system $R <n, m>$ in n **MA** time units, using m **MA** elements in a closed-loop structure.

8.4.3 Evaluation of Continued Fractions

Since continued fractions may be expressed recursively, we would expect to find application here also for our recursive evaluator. For instance, consider

$$F = \cfrac{1}{a_1 + \cfrac{1}{a_2 + \cfrac{1}{a_3 + \cfrac{1}{a_4 + \cfrac{1}{a_5}}}}} \qquad (8.24)$$

Now let $F_5 = 1/a_5$ and $P_5 = a_5$. Then

$$F_4 = \frac{1}{a_4 + F_5} = \frac{P_5}{P_4} \qquad \text{where } P_4 = 1 + a_4 P_5$$

$$F_3 = \frac{1}{a_3 + F_4} = \frac{P_4}{P_3} \quad \text{where } P_3 = P_5 + a_3 P_4$$

$$F_2 = \frac{1}{a_2 + F_3} = \frac{P_3}{P_2} \quad \text{where } P_2 = P_4 + a_2 P_3$$

$$F = F_1 = \frac{1}{a_1 + F_2} = \frac{P_2}{P_1} \quad \text{where } P_1 = P_3 + a_1 P_2$$

More generally,

$$P_n = a_n$$

$$P_{n-1} = 1 + a_{n-1} P_n$$

$$\vdots$$

$$P_j = P_{j+2} + a_j P_{j+1}$$

Finally, F may be evaluated by a single division step: $F = P_2/P_1$.

A chained structure of **MA** elements to calculate the P terms is shown in Fig. 8.18. (Since the feedforward paths skip over an intervening stage, we would require latches or delay elements to hold these signals until the product inputs were applied.)

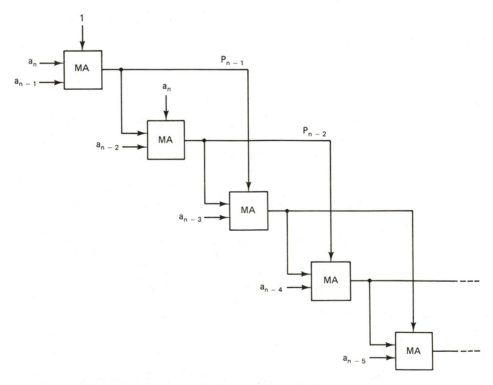

Figure 8.18 Continued fraction evaluation with iterated MA elements.

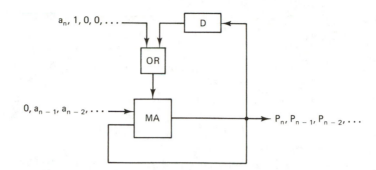

Figure 8.19   Continued fraction evalutation with a closed loop.

This chained circuit also has a closed-loop analog, as shown in Fig. 8.19. Alternatively, we may approximate F successively by

$$F^{(1)} = \frac{P_1}{Q_1} = \frac{1}{a_1} \qquad\qquad \text{where } P_1 = 1 \text{ and } Q_1 = a_1$$

$$F^{(2)} = \frac{P_2}{Q_2} = \cfrac{1}{a_1 + \cfrac{1}{a_2}} \qquad\qquad \text{where } P_2 = a_a \text{ and } A_2 = 1 + Q_1 a_2$$

$$F^{(3)} = \frac{P_3}{Q_3} = \cfrac{1}{a_1 + \cfrac{1}{a_2 + \cfrac{1}{a_3}}} \qquad\qquad \text{where } P_3 = P_1 + P_2 a_3 \text{ and } Q_3 = A_1 + Q_2 a_3$$

which gives rise to the well-known recursions

$$P_j = P_{j-2} + A_j \times P_{j-1} \quad \text{and} \quad Q_j = Q_{j-2} + A_j \times Q_{j-1}$$

with the initiating values

$$\begin{pmatrix} P_0 = 0 \\ P_1 = 1 \end{pmatrix} \quad \text{and} \quad \begin{pmatrix} Q_0 = 1 \\ Q_1 = a_1 \end{pmatrix}$$

As Robertson and Trivedi [21] point out, these recursions may be separately and simultaneously determined. It is clear that two recursive loops like that of Fig. 8.19 can be used to carry out these computations (with initializing inputs properly chosen for each).

It should be noted that the difference between the successive approximations $F^{(1)}, F^{(2)}, \ldots$, and the fractions $F_5, F_4, \ldots$, results from starting in the one case from the top of the continued fraction and in the other from the bottom. Starting from the bottom is possible only if one knows where the "bottom" is, and in the case of an infinite continuing fraction, a decision is required a priori concerning the point to designate as "bottom". Although proceeding from the top involves two recursive loops instead of one, it is probably preferable since the point of terminating the evaluation can be made to be a function of the convergence of the result.

8.4.4 Considerations in Construction of the MA Element

The most obvious problem in the design of the fixed-point MA element is the handling of overflows. When we form $W = X \times Y + Z$, where X, Y, and Z are fractions, $X \times Y$ will be a fraction, but the result W can be almost as large as 2. Clearly, in a sequence of MA operations, such overflows could continue to accumulate, so that for an nth-degree polynomial we might find the result to be almost as large as $(n + 1)$. We can, if we wish, regard overflows as nonvalid cases and require that problems be scaled so as to avoid overflow in the intermediate steps and in the result. However, since networks of MA elements are more likely to be used in special-purpose computing systems where something is known from the beginning about the ranges of the variables to be encountered, it becomes practical to build in some guard digits, that is, digits to the left of the radix point to accommodate the maximum range of overflow that will occur.

In a chain of MA elements to evaluate a polynomial, as shown in Fig. 8.8, $p_1(x)$ can be nearly as large as 2, and therefore the first MA element should carry one binary digit in its integer part (as well as the sign bit). The next two elements can carry two binary digits in their integer parts, since their outputs will be smaller than 4. The next four elements need three binary digits in their integer part, and so forth, to whatever maximum degree of polynomial is to be evaluated.

If the reentrant network of Fig. 8.9 is used, the single MA element must carry enough integer digits to accommodate the final result. If the tree structure of Fig. 8.10 is used, the elements in the top row need only one guard digit, the elements in the second row must have two, and the elements in the general jth level must have j guard digits in their integer parts.

A particularly appropriate logic structure for the fixed-point multiply-add element is the Wallace tree (or the Dadda variant), described in Chapter 4. For most operand bit lengths, the addition of the vector Z to the vectors summed in forming $X \times Y$ can be accomplished without increasing the number of levels in the tree and therefore without incurring any delay beyond the time required for the multiplication alone. For example, a 16-bit MA element requires 16 ICAND vectors, a row of possible C_0 bits (to convert 1's-complement numbers to 2's-complement form, if the ICAND vector is subtracted), and a row for the Z vector. These 18 vectors can be summed to two vectors by six levels of the Wallace tree, the same as the number of levels needed for only the multiplication. Indeed, as Dadda showed, any number of vectors from 14 through 19 can be summed in six levels, any number from 20 through 28 in seven levels, and so on. If the inclusion of the summand vector Z does not increase the number of vectors into the next higher group, no extra time is required for the addition of Z.

In principle, it is also possible to design a floating-point MA element, although the structure will be much more complex than the fixed-point structure. Provision must be made for the usual pre- and post-alignment steps and for exponent overflow and underflow. Futhermore, since the position in which the Z vector is to be added to the product $X \times Y$ depends on the exponents of these components, the addition of Z cannot be as readily incorporated into the Wallace tree as with the fixed-point structure, and thus a possible speed advantage is lost.

8.5 ON-LINE COMPUTATION

An interesting and potentially very useful method for doing arithmetic operations and function evaluation has been introduced by M. D. Ercegovac [22] and substantially expanded by him and by others [23–27]. In this method operands are presented serially, digit by digit, starting usually with the most significant, and result digits are also formed serially at the same rate as the operand digits but with a time delay. The method is made possible by representing the operands in a redundant digit form that limits the propagation of carries; otherwise, the most signficant digits of the result could not be determined until all operand digits have been presented and any possible carries have had time to propagate fully.

Conventional addition and subtraction can be done "on-line" if the digits are presented from right to left, that is, starting with the least significant digits. We saw an example in Fig. 3.8, where binary operands were applied a digit at a time and the binary result appeared sequentially, at the same rate as the input digits but delayed from them by the time delay of the adder. Such a structure is called a *fully digit on-line* structure, since both of the operands and the result are sequential digit by digit. The on-line property need not hold both for operands and result—division, for instance, is ordinarily on-line with respect to the quotient digits, but all of the divisor digits and as many dividend digits must be furnished before any quotient digits are formed. The most useful on-line procedures, however, are those that are fully on-line and which, furthermore, start from the left end. Left-to-right on-line procedures are compatible with digital information obtained from analog-to-digital converters, in which the most significant digits are formed before the less significant digits.

To describe the procedure for on-line multiplication, we will follow the original method presented by Trivedi and Ercegovac [23]. We consider two m-digit positive fractions:

$$X = \sum_{i}^{m} x_i \times r^{-i}$$

$$Y = \sum_{1}^{m} y_i \times r^{-i}$$

The leftmost j digits of these groups are denoted X_j and Y_j, where

$$X_j = \sum_{1}^{j} x_i \times r^{-i} = X_{j-1} + x_j \times r^{-j}$$

$$Y_j = \sum_{1}^{j} y_i \times r^{-i} = Y_{j-1} + y_j \times r^{-j}$$

The partial product obtained by multiplying these two j-digit groups is

$$X_j \times Y_j = X_{j-1} \times Y_{j-1} + (X_j \times y_j + Y_{j-1} \times x_j) \times r^j$$

This jth partial product, scaled by the factor r^j, is

$$P_j = X_j \times Y_j \times r^j$$

which may be written in recursive form as

$$P_j = r \times P_{j-1} + X_j \times y_j + Y_{j-1} \times x_j$$

where we take $P_0 = 0$ and $Y_0 = 0$.

This is an on-line procedure with respect to the two operands, but if we use non-redundant numbers, it does not generate the product digits sequentially, as may be readily verified by using it to form (for example) the product of 0.101×0.0101. The redundant number set that we will use is a symmetric one,

$$D = \{-\rho, -(\rho - 1) \cdots, -1, 0, 1, \cdots, (\rho - 1), \rho\}$$

where we take ρ to be

$$\frac{r}{2} \le \rho \le r - 1$$

with r assumed to be an even number.

We now define the product digits as d_j and write a related recursion

$$w_j = r(w_{j-1} - d_{j-1}) + X_j \times y_j + Y_{j-1} \times x_j$$

which leads to

$$w_j = P_j - \sum_{i=1}^{j-1} d_i \times r^{(j-i)}$$

The individual digits d_j, chosen from the balanced redundant set D, are found by

$$d_j = (\text{sign of } w_j) \times \lfloor |w_j| + 1/2 \rfloor$$

Finally, setting $j = m$, we obtain

$$w_m = P_m - \sum_{1}^{m-1} d_i \times r^{(m-i)}$$

$$P_m = X_m \times Y_m \times r^m$$

$$X \times Y = \sum_{1}^{m} d_i \times r^{-i} + (w_m - d_m) \times r^{-m}$$

Note that m iterations of this algorithm form the first m digits of the product, one after the other. The last m digits of the product are comprised by the group $(w_m - d_m) \times r^{-m}$. Of course, the resulting product is in redundant form, and a simple conversion step is needed to restore it to conventional nonredundant form.

We should also observe that since the digit d_j must be one of the elements of D,

TABLE 8.3

j	$X_j y_j + Y_{j-1} x_j$	w_j	d_j	$2 \times (w_j - d_j)$
1	0	0	0	0
2	0.01	0.01	0	0.1
3	0.101	1.001	1	0.01
4	0.0110	0.1010	1	-0.11
5	0.0111	-0.0101	0	-0.101
6	0	-0.101	-1	0.11
7	0.0110100	1.0010100	1	0.010100
8	0.11011011	1.00101011	1	0.0101011

we must have each $|d_j| \leq \rho$. The rule for determining d_j, together with this condition, requires that

$$|w_j| < \rho + 1/2$$

for all j. This, in turn, sets bounds on the magnitudes of the operands, so that for $r = 2$, we require $|X|$ and $|Y| \leq 1/2$.

Table 8.3 (from Ref. 23) tabulates the successive iterates in the multiplication of $X = 0.01101001$ by $Y = 0.01110011$ by this method. The first eight digits of the product are thus 0.00110111 which converts to 0.00101111 in nonredundant form. To these we append $(w_8 - d_8) \times 2^{-8}$ to get the 16-bit result $X \times Y = 0.0010111100101011$.

Similar procedures have also been defined for division and for square root. Application of this technique to floating-point arithmetic has also been studied, although floating-point is complicated by the separate treatment needed for exponents and for significands.

REFERENCES

1. Chen, T. C., Automatic computation of exponentials, logarithms, ratios and square roots, *IBM Jl. R&D*, vol. 16, no. 4, July 1972, pp. 380–388.

2. Meggitt, J. E., Pseudo division and pseudo multiplication processes, *IBM Jl. R&D*, vol. 6, no. 2, Apr. 1962, pp. 210–226.

3. Volder, J. E., The CORDIC trignometric computing technique, *IRE Trans. El. Comp.*, vol. EC-8, no. 3, Sept. 1959, pp. 330–334.

4. Haviland, G. L., and A. A. Tuszynski, A CORDIC arithmetic processor chip, *IEEE Trans. Comp.*, vol C-29, no. 2, Feb. 1980, pp. 68–79.

5. Daggett, D. H., Decimal–binary conversion in CORDIC, *IRE Trans. El. Comp.*, vol. EC-8, no. 3, Sept. 1959, pp. 335–339.

6. Steer, D. G., and S. R. Penstone, Digitial hardware for sine–cosine function, *IEEE Trans. Comp.*, vol. C-26, no. 12, Dec. 1977, pp. 1283–1286.

7. Walther, J. S., A unified algorithm for elementary functions, *Proc. AFIPS Spring Joint Comp. Conf.*, 1971, pp. 379–385.

8. Linhardt, R. D., and H. S. Miiller, Digit-by-digit transcendental-function evaluation, *RCA Rev.*, vol. 30, 1969, pp. 209–247.

9. Wrathall, C., and T. C. Chen, Convergence guarantee and improvements for a hardware exponential and logarithm evaluation scheme, *Proc. 4th Symp. on Comp. Arith.*, Oct. 1978, IEEE Cat. no. 78CH1412-6C, pp. 175–182.

10. Specker, W. H., A class of algorithms for $\ln X$, $\exp X$, $\sin X$, $\cos X$, $\tan^{-1} X$, and $\cot^{-1} X$, *IEEE Trans. El. Comp.,* vol. EC-14, no. 1, Feb. 1965, pp. 85–86.

11. Cochran, W. T., et al., What is the fast Fourier transform?, *Proc. IEEE,* vol. 55, no. 10, Oct. 1967, pp. 1664–1674.

12. Bowen, B. A., and W. R. Brown, *VLSI Systems Design for Digital Signal Processing*, vol. 1, Prentice-Hall, Englewood Cliffs, N. J., 1982.

13. Karplus, W. J., and D. Cohen, Architectural and software issues in the design and application of peripheral array processors, *Computer*, vol. 14, no. 9, Sept. 1981, pp. 11–17. This issue of *Computer* also contains several other papers on array processors.

14. Pease, M. C., An adaptation of the fast Fourier transform for parallel processing, *Jl. ACM*, vol. 15, no. 2, Apr. 1968, pp. 252–264.

15. Pease, M. C., Organization of large scale Fourier processors, *Jl. ACM*, vol. 16, no. 3, July 1969, pp. 474–482.

16. Gold, B., and T. Bially, Parallelism in fast Fourier transform hardware, *IEEE Trans. Audio Electroacoust.*, vol. AU-21, no. 1, Feb. 1973, pp. 5–16.

17. Liu, B., and A. Peled, A new hardware realization of high-speed fast Fourier transformers, *IEEE Trans. Acoust., Speech, Signal Process.*, vol. ASSP-23, no. 6, Dec. 1975, pp. 543–547.

18. Despain, A. M., Very fast Fourier transform algorithms hardware for implementation, *IEEE Trans. Comp.*, vol. C-28, no. 5, May 1979, pp. 333–341.

19. Glass, J. M., An efficient method for improving reliability of a pipeline FFT, *IEEE Trans. Comp.*, vol. C-29, no. 11, Nov. 1980, pp. 1017–1020.

20. Estrin, G., Organization of computer systems—the fixed plus variable structure computer, *Proc. 1960 Western Joint Comp. Conf.*, pp. 33–40.

21. Robertson, J. E., and K. Trivedi, The status of investigations into computer hardware design based on the use of continued fractions, *IEEE Trans. Comp.* vol. C-22, no. 6, June 1973, pp. 555–560.

22. Ercegovac, M. D., A general hardware-oriented method for evaluation of functions and computations in a digital computer, *IEEE Trans. Comp.*, vol. C-26, no. 7, July 1977, pp. 667–680.

23. Trivedi, K. S., and M. D. Ercegovac, On-line algorithms for division and multiplication, *IEEE Trans. Comp.*, vol. C-26, no. 7, July 1977, pp. 681–687.

24. Irwin, M. J., and R. M. Owens, Fully digit on-line methods, *IEEE Trans. Comp.*, vol. C-32, no. 4, Apr. 1983, pp. 402–406.

25. Ercegovac, M. D., and A. L. Grnarov, On the performance of on-line arithmetic, *Proc. 1980 Conf. on Parallel Process.*, pp. 55–62.

26. Gorji-Sinaki, A., and M. D. Ercegovac, Design of a digit-slice on-line arithmetic unit, *Proc. 5th Symp. on Comp. Arith.*, May 1981, pp. 72–80.

27. Oklobdzija, V. G., and M. D. Ercegovac, An on-line square root algorithm, *IEEE Trans. Comp.*, vol. C-31, no. 1, Jan. 1982, pp. 70–75.

Index

Proper Names

Subject